HERE'S NESSIE!

A Monstrous Compendium from Loch Ness

Dr Karl P.N. Shuker

Typeset by Jonathan Downes, Charlotte Phillipson, Chloe Gray
Cover and Layout by SPiderKaT for CFZ Communications
Using Microsoft Word 2000, Microsoft Publisher 2000, Adobe Photoshop CS.

First published in Great Britain by CFZ Press

CFZ Press
Myrtle Cottage
Woolsery
Bideford
North Devon
EX39 5QR

ISBN: 978-1-909488-45-8

DEDICATION

In memory of Tim Dinsdale (1924-1987) – Nessie's foremost investigator, who was also the very first cryptozoologist to praise and encourage my own fledgling researches and writings concerning mystery beasts way back in the mid-1980s. A seeker of elusive animals, he was himself that rarest of creatures – a true gentleman. The world is a much more prosaic, much less courteous place without him.

Tim Dinsdale in 1979 (© Prof. Henry H. Bauer)

LIST ⊕F C⊕NTENTS

ACKN⊕WLEDGEⅢENTS

As always, many people contributed in many ways to the preparation of this book, but I wish to offer my especial thanks to the following persons:

Neil Arnold, Janet Bord/Fortean Picture Library, Markus Bühler, the late Mark Chorvinsky, Karl J. Claridge, Jay Cooney/*Bizarre Zoology*, Jane Cooper, Mike Dash, Claudio Diaz, the late Tim Dinsdale, Robert Elsmore, Thomas Finley, Errol Fuller, Rip Hepple/*Nessletter*, Jeff Johnson, Connor Lachmanec, Peter Loh, the late Prof. Roy P. Mackal, Ulrich Magin, Joe Mancini, Scott Mardis, Jeff Meuse, Tim Morris, Wm Michael Mott, Richard Muirhead, Adam Naworal, Andy Paciorek, Michael Playfair, Richard Pullen, Jeff Rausch, William M. Rebsamen, the late Dr Robert H. Rines/Academy of Applied Science, Tony 'Doc' Shiels, Lance Shirley, the late Mary D. Shuker, Paul Sieveking, Richard Svensson, Leslie Thomson, Anthony Wallis, Dave Walsh, Roland Watson, Rod Williams, Chris Woodyard, Valerie Wyllie.

In addition, I am as always greatly appreciative of the interest and enthusiasm shown by my publisher Jonathan Downes at CFZ Press; I am exceedingly grateful to longstanding LNM researcher Prof. Henry H. Bauer, author of the classic scholarly work *The Enigma of Loch Ness*, for so kindly contributing his delightful foreword to this present book of mine; and I owe a massive debt of thanks to inspirational artists Wm Michael Mott and Anthony Wallis for their wonderful cover artwork.

SOURCES OF THIS BOOK'S CHAPTERS

This book is a compendium volume of my previously-published writings on the subject of the Loch Ness monster. All of its chapters are reprinted, expanded, updated, or amalgamated versions of text drawn from my following books, articles, and *ShukerNature* blog posts:

In Search of Prehistoric Survivors: Do Giant 'Extinct' Creatures Still Exist? (Blandford Press: London, 1995).

'The Long-Necked Seal in Cryptozoology – Part 1: Giraffe Seals and Sea Serpents' (*ShukerNature*, http://www.karlshuker.blogspot.co.uk/2015/07/the-long-necked-seal-in-cryptozoology.html 28 July 2015).

'The Long-Necked Seal in Cryptozoology – Part 2: From Swan-Necked and Hidden-Necked to Tizheruk and Nessie' (*ShukerNature*, http://www.karlshuker.blogspot.co.uk/2015/07/the-long-necked-seal-in-cryptozoology_29.html 29 July 2015).

'Dating Nessie' [in my Menagerie of Mystery column] (*Strange Magazine*, no. 20, December 1998), pp. 38-39.

'The 'Other' Encounter Between a British Saint and a British Lake Monster' (*ShukerNature*, http://karlshuker.blogspot.co.uk/2014/07/the-other-encounter-between-british.html 4 July 2014).

'In Conference With Cryptozoology' (*The Unknown*, no. 28, October 1987), pp. 24-29.

'What May Be a Hitherto-Undocumented Sighting of the Loch Ness Monster, Given To Me by Tim Dinsdale on 25 July 1987' (*ShukerNature*, http://karlshuker.blogspot.co.uk/2013/07/what-may-be-hitherto-undocumented.html 30 July 2013).

'Never Bother a Beithir' (*Wild About Animals*, vol. 8, July 1996), p. 9.

'Is Nessie a Plesi or Something Even Stranger?' (*Paranormal Magazine*, no. 32, February

2009), pp. 38-43.

'Does the Loch Ness Monster Have a Split Personality? Revealing Nessie's Strangest Identities' (*ShukerNature*, http://karlshuker.blogspot.co.uk/2013/07/does-loch-ness-monster-have-split.html 16 July 2013).

'A Frog as Big as a Goat – The Most Mystifying Loch Ness Monster Sighting Ever?' (*ShukerNature*, http://karlshuker.blogspot.co.uk/2016/05/a-frog-as-big-as-goat-most-mystifying.html 10 May 2016).

Extraordinary Animals Revisited: From Singing Dogs To Serpent Kings (CFZ Press: Bideford, 2007).

Karl Shuker's Alien Zoo: From the Pages of Fortean Times (CFZ Press: Bideford, 2010).

'It Ain't Nessie-ssarily So - Or Is It? [review of the Discovery Channel DVD *Loch Ness Discovered*] (*ShukerNature*, http://karlshuker.blogspot.co.uk/2009/02/it-aint-nessie-ssarily-so-or-is-it.html 25 February 2009).

'*Loch Ness, Nessie, & Me. The Truth Revealed* by Tony Harmsworth' [my review of this book] (*Journal of Scientific Exploration*, vol. 25, summer 2011), pp. 404-406.

'In Tune With Nessie – A Selection of Loch Ness Monster-Themed Songs' (*ShukerNature*, http://karlshuker.blogspot.co.uk/2015/09/in-tune-with-nessie-selection-of-loch.html 23 September 2015).

Star Steeds and Other Dreams: The Collected Poems (CFZ Press: Bideford, 2009).

'A Nessie-ssary Rhyme!' (*ShukerNature*, http://karlshuker.blogspot.co.uk/2011/01/nessie-ssary-rhyme.html 16 January 2011).

'Taking a Shine to Nessie Receives the Stamp of Approval – The LNM on a FDC' (*ShukerNature*, http://karlshuker.blogspot.co.uk/2013/08/taking-shine-to-nessie-receives-stamp.html 7 August 2013).

'Keeping Clear of Kelpies and Other Water-Horses' (*ShukerNature*, http://karlshuker.blogspot.co.uk/2015/04/keeping-clear-of-kelpies-and-other.html 23 April 2015).

'Pony...or Pooka?' (*ShukerNature*, http://www.karlshuker.blogspot.co.uk/2010/12/ponyor-pooka.html 30 December 2010).

Mysteries of Planet Earth: An Encyclopedia of the Inexplicable (Carlton Books: London, 1999).

'Lesser Nessies – Surveying the 'Other' Monsters of Mainland Scotland's Freshwater

Lochs' (*ShukerNature*, http://karlshuker.blogspot.co.uk/2015/07/lesser-nessies-reviewing-other-monsters.html 13 July 2015).

'White Mice, Bumblebees, and Alien Worms? Unexpected Mini-Monsterlings in Loch Ness' (*ShukerNature*, http://karlshuker.blogspot.co.uk/2016/03/white-mice-bumblebees-and-alien-worms.html 28 March 2016).

'In Search of the Missing Monster' (*Fortean Times*, no. 253, September 2009), pp. 52-55.

'Loch Watten's Missing Monster – Whatever Happened to Wattie?' (*ShukerNature*, http://karlshuker.blogspot.co.uk/2012/06/loch-wattens-missing-monster-whatever.html 11 June 2012).

'When Trunko Met Nessie?? - Paradox of the Pictish Beast' (*ShukerNature*, 13 September 2015).

'The Irish Master Otter in Scotland?' (*ShukerNature*, http://karlshuker.blogspot.co.uk/2011/02/irish-master-otter-in-scotland.html 20 February 2011).

'The First Nessie Book, and a Comprehensive Bibliography of the Loch Ness Monster' (*ShukerNature*, http://karlshuker.blogspot.co.uk/2014/06/the-first-nessie-book-and-comprehensive.html 25 June 2014.

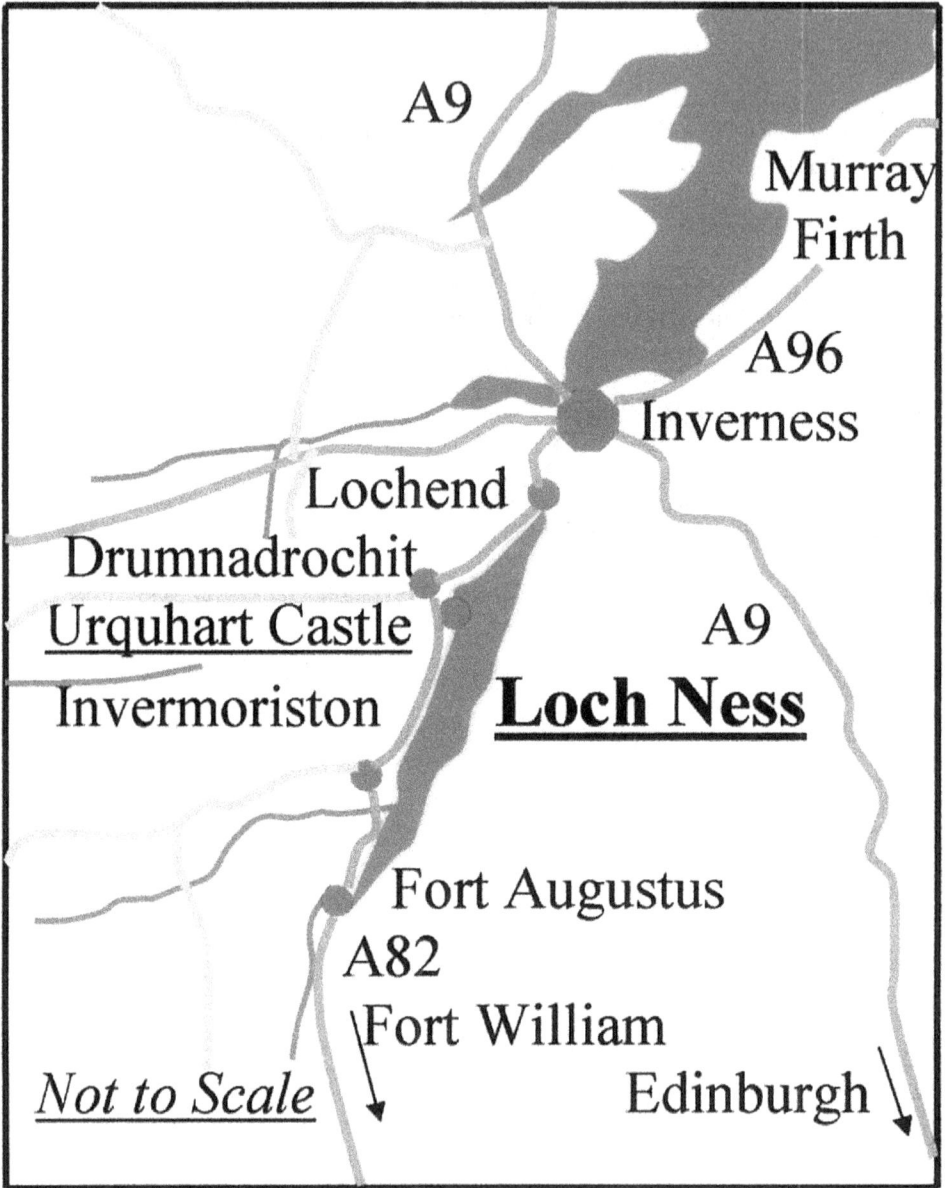

A9

Murray
Firth

A96
Inverness

Lochend

Drumnadrochit

Urquhart Castle

A9

Invermoriston

Loch Ness

Fort Augustus

A82

Fort William

Not to Scale

Edinburgh

Sketch-map of Loch Ness and its environs (public domain)

FOREWORD BY PROF. HENRY H. BAUER
AUTHOR OF: *THE ENIGMA OF LOCH NESS: MAKING SENSE OF A MYSTERY*

This compilation about Loch Ness Monsters is wonderfully diverse, comprehensive, and well chosen, and it exemplifies Karl Shuker's enormous range of zoological knowledge, familiarity with relevant literature, and ability to provide fresh and thought-provoking analysis. One highlight is the well documented discussion in Chapter 1 of the various mutually contradictory claims that the iconic Surgeon's photo was faked, a hoax. A comprehensive discussion of the possible relation of Nessies to plesiosaurs is also very illuminating, for instance about what evolution could have brought during the 66-million gap in the fossil record — I learned something I had seen nowhere else, that today's coelacanths are 5 times larger than many of their fossil ancestors were; and that sufficiently large reptiles "lose body heat so slowly that to all intent and purposes they are functional endotherms [which generate their own body heat]"; further, that great enough activity can also generate heat even in reptiles. A very important point, not widely enough known, is that the estimated mass of fish in Loch Ness was set in 1993 at 27 tons, ample enough for a family of Nessies, which the 1973 guess of 9 tons had not been. But this book goes far beyond the scientific stuff, with songs and pictures and poems illustrating that Nessies are part of modern culture everywhere; there was even a stamp issued by the Maldives showing a Nessie-like neck and head as iconic of "Mysteries of the Universe".

I have enjoyed and benefited from everything I've read of Karl Shuker's work. In my view this book exceeds all the rest. One reason for that is rather personal, the other extends that to what I believe to be quite general.

(© Prof. Henry H. Bauer)

The personal reason is that I treasure the same memory of and enormous regard for Tim Dinsdale as in this book's Dedication. Karl mentions later that one of Tim's books, *The Story of the Loch Ness Monster*, had hooked him; and that was true for me too, albeit it was Tim's first one, *Loch Ness Monster*.

My more general reason is the belief that, for both Karl and me, getting hooked has brought all sorts of good things over the years, and I believe firmly that similarly good things are likely to happen for everyone who gets hooked on something and pursues it faithfully and wherever it leads. It doesn't matter what the specific thing happens to be that leads to determined and open-ended inquiry, it's that the process of trying to get at the truth about something, even an apparently limited, specific subject, leads almost inevitably to having to learn about all sorts of other things, and likely to becoming interested in many other matters, and making all sorts of contacts with people from whom one can learn much.

I didn't realize it at the time, of course, but in retrospect I understand that becoming intensely curious about Tim Dinsdale's film and his belief in Nessies' reality did at least as much for me as my formal education had done, by making me try ever harder to think critically, to seek out potential evidence and examine it as closely and objectively and scrupulously as possible, and above all to be critical about "what everyone knows", what the recognized experts and authorities say. Some of those experts and authorities denigrate interest in Nessies and other such matters as pseudo-science, but through my quest for the truth about Nessies I learned a great deal about what science really is, and how misleading so many common ideas about science are. That I had been hooked by Tim's book in about 1961 led to a major career change in the late 1970s that has brought enormous intellectual (as well as tangible) rewards, among them a thankfully rather short administrative stint as a Dean of Arts & Sciences, so that I could on occasion introduce Tim Dinsdale to an audience as "not only a monster hunter but also a Dean-maker"; and the first book I wrote that was not in my original professional specialty of chemistry is dedicated to "Tim Dinsdale, who planted the seed".

Since it remains mysterious, what Nessies really are, this quest can still serve as an excellent thing to get hooked on; but it could equally be one of the other cryptids and curiosities that Karl has discussed elsewhere; for instance, the nature of "the Pictish beast" described in Chapter 15 of this Nessie compilation is a riddle that cries out for a compelling explanation. Many people also become fascinated by one or other of the unexplained matters that fascinate various of my colleagues in the Society for Scientific Exploration: psychic phenomena (which, incidentally, also intrigued Tim Dinsdale), UFOs, or even who the actual author was of the plays and poems officially attributed to William Shakespeare.

At any rate, for those already hooked on Nessies as well as for those who are not yet hooked, this compilation of pertinent material forms an available treasure trove.

INTR⊕DUCTI⊕N

The young couple were in a happy mood that morning – and why not? It was summer, the war was just over, and they were off on a holiday. Taking the lakeside road they headed the car north, enjoying the scenery and a sense of freedom...until...bang went one of their tyres! While Mr. Atkinson changed the wheel his wife admired the beauty of the great lake.

It was breath-taking. All the majestic grandeur and calm of unspoiled nature. Abruptly, the peace was shattered, and her blood ran cold. The placid lake waters boiled and heaved, suddenly, and up from the depths came a great reptilian neck followed by a huge humped body, long, grey, and powerful...

It was terrifying and repulsive...She called frantically to her husband. Hardly able to believe their eyes, but equally unable to doubt them, they both watched the monster until it sank out of sight...Years later Mrs. Atkinson still could not think of that day without turning cold. She would remember the terrifying sight until the end of her days...

This uniquely grim incident took place in 1946 [early August] and it was related later to a scientist enquiring into the mystery of the Loch Ness Monster [Dr Maurice Burton – see below]. Why should this story be singled out among so many?...

A leading scientist [Burton] has now admitted that he feels bound to regard this [Norah Atkinson's sighting] as important 'because it has every indication of being genuine'. Mrs. Atkinson could not have been misled by freak wave or light effects, otters, floating vegetable matter, or any of the things usually suggested as an alternative explanation. She was too close to the bank, it was a clear day, her impression was too definite, and her husband confirmed it. Both of them had always been sceptical about reports of the monster. Yet, they abandoned their holiday journey, and drove straight back to Inverness to tell relatives what they had seen. To this day Mrs. Atkinson maintains she will never be able to forget that awesome sight: "Goodness knows what else the loch hides," she was reported as saying. "The very thought makes me turn cold."

'What lives in Loch Ness?', in *Stranger Than People*

[In summer 1959, Norah Atkinson described her sighting firsthand in a letter written by her to British zoologist and Nessie investigator Dr Maurice Burton, who included it in his book *The Elusive Monster*, published in 1961. In her letter, Atkinson's verbatim description of Nessie was: "Suddenly, there was a terrific upheaval of water and up came a long swan-like neck with a small head. Then the whole body appeared, elephant grey in colour, two humps and very long and powerful...We both watched until it disappeared underneath the water just below the edge of the bank. What a terrifying sight it was."]

It is all thanks to my mother Mary Shuker taking me as a child to the town of Walsall in the West Midlands, England, one afternoon during the late 1960s that I first learnt of the Loch Ness monster (LNM). For it was while we were browsing together that day in Walsall's branch of the books and stationery shop W.H. Smith that Mom bought me a thoroughly amazing, captivating book entitled *Stranger Than People*.

Published in 1968, it was packed with lavishly-illustrated spreads documenting mysteries from the real world, the world of myths and legends, and the world of fiction (which included some wonderful, specially-written short stories). It is within the pages of this momentous, life-changing volume, one that profoundly influenced and nurtured what has become my lifelong interest in mysterious phenomena (especially of the cryptozoological and zoomythological kinds), that I first read about such diverse but fascinating subjects as the yeti, the colossus of Rhodes, Edgar Cayce and telepathy, werewolves and vampires, the kraken, giants, the minotaur, Von Kempelen's chess player, mermaids, witches and witchcraft, 'deathless' warriors, aliens, trolls, zombies, leprechauns, feral children, Herne the hunter, Moby Dick – and, of course, Nessie, the Loch Ness monster (LNM), from whose chapter in that book the quote opening this introduction to my own present book is excerpted.

A few years later, moreover, during a visit to the Warwickshire, England, town of Stratford-upon-Avon (birthplace of William Shakespeare), Mom bought me my first LNM book, Tim Dinsdale's *The Story of the Loch Ness Monster* (1973). An enthralling account of this Scottish mystery beast's remarkable history, it was justifiably lauded as a Commended Title in the prestigious Times Educational Supplement Information Book Awards during that same year.

Needless to say, I was hooked - and have been in relation to this long-necked aquatic cryptid ever since – unlike Nessie, conversely, which continues to be just as evasive today as way back then, over 40 years ago now, but, equally, remains just as thought-provoking and enigmatic too.

With two wonderful, life-changing books – *The Story of the Loch Ness Monster* and *Stranger Than People*; and with one wonderful, life-changing lady – my mother, Mary Shuker (© Dr Karl Shuker)

Indeed, the LNM is not only the premier mystery beast of the United Kingdom, it also vies with the bigfoot or sasquatch as the most famous one anywhere in the world. Little wonder, therefore, that during my many years as a cryptozoological researcher and writer I should have documented it and all manner of aspects relating to it in a wide range of publications.

Now, however, for the very first time and in direct response to popular demand, the vast majority of these previously disparate Nessie-themed writings of mine have finally been brought together, and in expanded, updated form whenever possible too, to yield the present LNM compendium, covering a fascinating, extremely broad spectrum of pertinent topics.

The first two chapters set the scene by offering a comprehensive review of the history and controversies associated with this most contentious of aquatic cryptids. The others that follow them investigate such subjects as some of the most – and least – plausible taxonomic identities that have been proposed for it; the closely-linked traditional Scottish folklore of kelpies and other water-horses; a look at various Nessie-related hoaxes; an extensive survey of less familiar water monsters reported from other Highland lochs; some reviews of Nessie-related material; a hitherto-unpublished LNM sighting given to me by none other than leading Nessie researcher and eyewitness Tim Dinsdale shortly before his untimely death; Nessie and the noble hobby of philately; a retrospective devoted to the historic LNM conference staged by the International Society of Cryptozoology at Edinburgh's Royal Museum of Scotland in 1987; the abiding paradox of the mystifying Pictish beast as intricately carved upon numerous ancient symbol stones by Scotland's early 'painted people' the Picts; my very own tribute in verse to Nessie; an annotated, YouTube-linked listing of Nessie-themed songs and music videos; and much more too!

Supplementing these diverse subjects is an equally eclectic selection of illustrations - a dedicated Nessie gallery containing a dazzling array of spectacular full-colour LNM artwork, including a number of specially-commissioned, previously-unpublished examples - plus a wide range of text images, a very comprehensive bibliography of non-fiction Nessie books, a listing of current LNM-themed websites, and a detailed index.

So without further ado, welcome to the sometimes decidedly weird yet always totally wonderful world of Nessie - the mystifying but ever-memorable monster of Loch Ness.

[**Author's note:** Constituting Chapters 1-2 of the present compendium, this two-part review of the LNM's history and associated controversies is excerpted (and updated where necessary) from one of my earliest books, *In Search of Prehistoric Survivors* (1995), which was commissioned and published by Blandford. Their brief to me was to prepare a book documenting those cryptids for each of whom – as the book's title suggests - an animal type surviving (albeit in evolved form) from prehistoric times into the present day was at least one zoological identity that had been proposed by investigators, and in my coverage to concentrate predominantly upon that particular identity wherever possible. So that is what I did, but I didn't always personally agree with the prehistoric survivor identity in all cases. So please bear this in mind when reading the following two-chapter review. Please also bear in mind that since that book of mine was published, a full 21 years ago now, some of the photographic and sonar evidence discussed here have been the subject of substantial additional research, which in the opinion of certain investigators has cast varying degrees of doubt upon their validity. In relation to this, I recommend readers to access the following online articles:

http://www.lochnessinvestigation.com/flipper.html

http://blogs.scientificamerican.com/tetrapod-zoology/2013/07/10/photos-of-the-loch-ness-monster-revisited/

I also recommend readers to access the following online site containing Prof. Henry H. Bauer's comprehensive refutations of those criticisms:

http://henryhbauer.homestead.com/LochNessFacts.html

Thank you!]

CHAPTER 1:
REVIEWING THE LOCH NESS MONSTER
- PART 1: ITS HISTORY AND MYSTERY

The word 'monster' has a definitely unscientific aura, but that which is nameless must be called something, and the Loch Ness Monster, or the L.N.M., are terms by which it has been most widely known. Recently the thing has been referred to in newspaper reports as 'Nessie', but for an important zoological specimen such a name seems undignified.

The late Dom Cyril Dieckhoff of the Abbey, Fort Augustus, found on his pastoral visits, during which he conversed in Gaelic, that the Monster had long been known among local people as *an Niseag*. Actually this is the word 'Ness' in Gaelic with the feminine diminutive ending. 'Nessie' therefore is not such a bad rendering after all.

Constance Whyte – *More Than a Legend: The Story of the Loch Ness Monster*

The most famous mystery beast of any type, and the archetypal long-neck lake monster, is undeniably the elusive creature allegedly inhabiting Loch Ness, which is situated on the Great Glen Fault that runs southwards from Inverness to Fort William in the Scottish Highlands. Nessie's distinguished history supposedly dates back at least to the 6th Century AD, when a mysterious water beast was encountered by St Columba (though in the River Ness, not in Loch Ness, and at about 580 AD, rather than 565 AD as given in many other works – see Chapter 4).

Since 1933, which saw the creation of a new motoring road (the A82) overlooking the northern shoreline of this immense but hitherto-secluded lake, Loch Ness has hosted countless sightings of what many people believe to be huge but unremittingly reclusive water beasts with long necks, small heads, and humped backs - and prior to the A82's arrival, there had always been local beliefs in the existence within the loch of strange, sometimes frightening animals generally referred to as kelpies or water-horses (see Chapter 12).

Unless they are not resident here but migrate to and fro from the sea via the River Ness, or unless they habitually 'loch-hop' from one lake to another, if long-necks do occur in Loch Ness on a long-term basis there must be a viable population.

A view out over Loch Ness (© Dr Karl Shuker)

Loch Ness is approximately 23 miles long, up to 1.5 miles wide in places, has a maximum confirmed depth of 813 ft (but see later in this chapter for a remarkable unconfirmed claim made in January 2016), a volume of about 263 billion cubic feet (Scotland's biggest loch by volume, and containing more water than all of the lakes in England and Wales combined), a surface area of 22 square miles (Scotland's second-largest loch by surface area, beaten only by Loch Lomond), and a diverse biota with substantial quantities of fish readily able to sustain such a population, despite what critics and sceptics might claim (see Chapter 9 for exciting proof of this and also Nessie investigator Roland Watson's very thought-provoking calculations at: http://lochnessmystery.blogspot.co.uk/2012/02/is-there-enough-food-for-nessie_12.html). Consequently, problems of adequate space and food are easily dismissed - unlike the riddle of these beasts' identity.

As the Loch Ness monster's eventful history has been thoroughly examined in many full-length Nessie treatments currently in existence, this necessarily brief coverage will concentrate upon those aspects of the beast most relevant to its identification.

The most popular image of Nessie (or an Niseag in Scottish Gaelic, which apparently translates as 'pure') is that of a long slender neck and one or more humps rising briefly up above the loch's surface, thereby assigning it to the long-neck category of water monster – and which, with minor variations, has been reported by numerous eyewitnesses over the years, as shown by the following

**Diagram illustrating various observed forms attributed to Nessie
(public domain)**

selection of sightings (over 1000 alleged Nessie sightings have been recorded in total).

On 12 July 1934, naval officer Captain F.E.D. Haslefoot was walking along a disused railway line leading from Fort Augustus to Inchnacardoch Point when he saw two oval black humps in the loch, moving slowly towards the shore. They were roughly 8 ft apart, and 1 ft or so above the water surface, but as he watched they suddenly turned and travelled away up the loch for about 200 yards at a speed of about 10 knots, one in front of the other with the anterior hump generating a feather-like wash, before slowing down again to around 3 knots. As he watched, a head and neck emerged in front of one of the two humps, black in colour and roughly 4 ft out of the water. Haslefoot described the head as being spade-shaped, but only saw it for a moment before the creature flicked it sideways back down into the water. After swimming leisurely until about 50 yards from the loch's shore, it submerged.

In October 1936, from a spot about 3 miles outside Foyers, Marjory Moir and some friends enjoyed a 14-minute-long view of a long-necked beast with three humps, which was apparently resting on the water surface. It was estimated by them to be approximately 30 ft long, with humps of differing sizes (the middle one was largest, the anterior one smallest) and a slender neck whose head had no discernible features and frequently dipped back and forth into the water. Then, without warning, it sped away towards Urquhart Bay, throwing up a huge wash, before returning to its original resting site.

On 11 May 1962, from the window of a cottage at Alltsaigh, Edith Christie spied a single-humped long-neck with an egg-shaped head about 160 yards away, generating a tremendous wash as it swam rapidly northward up the loch at the speed of a fast motor boat. Its neck was about 4-5 ft long, and a 2-ft-long greenish-black hump followed about 6 ft behind it.

During a 12-minute sighting by Fort Augustus resident Katherine Robertson and her friend, a Lutherian nun visiting from Germany, two of these creatures were seen together, and for part of that time they were no more than 300 yards away. The date was 18 August 1971, and the two women were walking along the road behind Bolum Bay, when Robertson spied an extremely large animal swimming near to the shore, but away from the Abbey, on their left. The two friends observed it crossing the bay to the loch's far side, and Robertson estimated it to be about 45 ft long, with two low humps, and a small squarish head perched upon an erect, swan-like neck. As it was progressing across the bay, Robertson's friend spotted a second, smaller long-neck, also swimming from the vicinity of the Abbey and travelling quite speedily in the direction of the larger one until it finally caught up with it on the opposite side. This second animal had the same form of head and neck as the larger one, but only one hump was visible, and the length of the creature itself was estimated to be a mere 15 ft or so. After a few moments together, the two beasts submerged.

One morning in January 1980, while the loch's waters were totally still, fisherman Donald MacKinnon viewed a long-neck for almost 10 minutes with binoculars, noting three humps: "...and a long neck which reminded me of the submarine periscopes I'd seen in the war".

A surprising number of reports are on file concerning mysterious, unidentified submarines appearing in lakes and stretches of ocean throughout the world, whose presence has been made known only via brief glimpses of their periscopes projecting up above the water surface. Scandinavian fjords seem particularly prone to such appearances, and there has been much discussion regarding unexplained appearances and disappearances during World War II of craft assumed (but never proven) to have been periscope-bearing U-boats.

Two very different lines of thought have been aired as to their identity. One is that they are spy vessels, covertly observing while remaining largely unobserved themselves - except for their tell-tale periscopes. The other is that they are not submarines at all - but rather that they are long-necks, which makes MacKinnon's description particularly noteworthy.

Reports of long-necks in Loch Ness and re Nessie in general continue to the present day, as does international interest in this cryptozoological mega-star.

In June 1993, for example, Edna MacInnes and David McKay claimed to have watched a huge beast with light-brown body and giraffe-like head and neck for 10 minutes before it dived out of sight.

Seven years later, 'Loch Ness 2000' was a major new exhibition based at Drumnadrochit, dealing with the 'official' natural history of this famous lake, but inevitably featuring Nessie too. Designed by veteran Ness naturalist Adrian Shine, 'Loch Ness 2000' was formally opened in June 1999 by renowned explorer Sir Ranulph Fiennes.

During late March 2000, Swedish explorer Jan-Ove Sundberg led a week-long GUST (Global Underwater Search Team) survey at Loch Ness, dubbed 'Nessie 2000', using hydrophones day and night in the hope of recording sounds of the cryptozoological kind within its deep waters. The hydrophones had a range of 5 miles in ideal conditions, and unlike sonar equipment, which might disturb any creatures that may exist here, they do not create any disturbance when recording. The outcome of this survey was that some mysterious sounds were indeed recorded, including a series of pig-like grunting noises whose frequency of 741-751 Hz was comparable to sounds produced by various very big, known aquatic species such as killer whales, walruses, and elephant seals.

July 2001 marked the tenth anniversary of Nessie seeker Steve Feltham's vigil at the lochside of Loch Ness. For the past decade he had diligently albeit unsuccessfully scoured the loch's waters for a sign of its legendary inhabitant, and had lived throughout those ten years in nothing more luxurious than old library van. Now, however, he hoped to swap it for a double-decker bus, converting the bottom deck into a Loch Ness exhibition, and was already planning his next decade here.

In a British survey conducted during autumn 2006, over 2000 adults across the UK were asked to give their opinion as to Scotland's most famous figure, present or past. Obviously, the surveyors expected someone like Sean Connery, Robert Burns, Ewan McGregor, William Wallace, or Robert the Bruce to top the poll. But no, for although all of those names did indeed appear in the list they were all trounced by a most unexpected name – Nessie, the Loch Ness monster! One can only assume that a fair few of those questioned had cryptozoological leanings!

Nevertheless, due to the dearth in recent years of noteworthy reports Nessie seekers were beginning to fear that Scotland's most famous mystery beast was no more. In December 2015, however, Gary Campbell, Keeper of the Official Loch Ness Monster Sightings Register, was delighted to announce that 2015 had been a "vintage" year for such reports. Needless to say, many claimed sightings *are* reported each year, but as Campbell points out, the vast majority can be readily explained and are therefore eliminated from consideration for addition to the register, but in 2015 the register had formally accepted no fewer than five as being unexplained by normal phenomena, the most for 13 years.

In January 2016, Keith Stewart, a 43-year-old Loch Ness tour-boat operator, announced that using state-of-the-art 3-D sonar equipment, he had obtained a sonar reading at the loch revealing that it was 889 ft deep, thus surpassing the previous record of 813 ft measured at a spot called Edwards' Deep, and that he had returned to this location several times and verified his reading. Moreover, the site of this as-yet-independently-unconfirmed reading was a hitherto-undiscovered loch-bottom trench, situated about halfway between the Clansman Hotel and Drumnadrochit. Not surprisingly, the media soon seized upon the possibility of this deep trench being a potential hideaway for Nessie, suggesting that during sonar scans of the loch in the past, the monster(s) may have hidden

The Surgeon's Photograph (Fortean Picture Library)

in it to avoid detection. Of course, one has to ask how – assuming that such creatures even exist anyway – they would have known that the loch was being scanned during such procedures, unless they can detect sonar?

After all, cetaceans utilise a sonar-like echolocation mechanism in navigation and locating prey – emitting sound waves and being guided by the precise nature of their echoes. Also needing mention here is that any creature hiding in the trench for any lengthy period of time would either need to be able to hold its breath underwater for that period or be able to breathe directly in water. Stewart also claimed that two weeks earlier, he had obtained a sonar image of what looked like a long object with a hump, lying on the loch bed, but which wasn't there when he scanned the loch bed again later. If nothing else, should the trench's existence and depth be confirmed it will demonstrate that there are indeed sizeable surprises still awaiting disclosure in the depths of this most mysterious of freshwater lakes.

Having said that, a state-of-the-art underwater robot named Munin, operated by Norwegian company Konigsberg Maritime and conducting a very extensive two-week survey of the loch during April 2016, found no trace of any such trench. What it did find, however, at a depth of over 700 ft on the loch floor and thereby confirming its technical expertise in underwater exploration, was the huge, long-lost Nessie model that had been specially created for the 1970 British movie *The Private Life of Sherlock Holmes*, starring Sir Robert Stephens as Holmes and directed by Billy Wilder, Unfortunately, however, this spectacular model, complete with a tall vertical neck and head, abruptly sank to the bottom of the loch when it was taken out onto the water after Wilder had ill-advisedly requested that its two humps be removed (this apparently disrupted its balance, causing it to overturn and sink). See Chapter 8 for more details.

As for photos of Loch Ness long-necks, even the most striking examples have failed to sustain universal acceptance. In December 1975, a team from the Academy of Applied Science (based at Concord, New Hampshire), led by its founder Dr Robert H. Rines, released two underwater photos obtained during their research at the loch in June 1975 that incited a media sensation. One seemed to show the long neck and body of a plesiosaur-like beast, and the other resembled a close-up of a horned head (this was dubbed the 'gargoyle' photo – see this book's Nessie Full-Colour Gallery).

Some LNM authorities, including Nicholas Witchell, later opined that the 'gargoyle' may simply be a decayed tree stump that was located and eventually dredged up in October 1987 from the area of the loch bottom where the photo was taken. Leading on from this are suggestions from some researchers that the 'body and neck' photo merely depicts sediment swirling up from the loch bottom around the edge of the 'gargoyle' tree stump, whose features must have been concealed by shadow. Yet in the opinion of various others, such 'solutions' as these appear more contrived than convincing.

The greatest controversy, however, is reserved for the Loch Ness long-neck photo par excellence - the 'Surgeon's Photograph', whose image of a head and slender neck protruding up through the water is everyone's immediate mental picture of Nessie whenever the subject of Loch Ness is raised. The picture was one of two that Lieut.-Col. Robert K. Wilson (1899-

Painting of Nessie inspired by the 'gargoyle' photo (© Thomas Finley)

1969), a London gynaecologist, claimed to have snapped after he had seen something on the loch from a spot about 2 miles north of Invermoriston during a car journey along the A82 on the morning of 21 April 1934.

Sceptics have variously attempted to dismiss the photo's 'head-and-neck' image as that of a bird, or the tail of a diving otter, or even as the dorsal fin of an out-of-place killer whale or of a giant sturgeon. During the 1990s, however, two rival allegations were made that sought to expose the photo as a blatant hoax - but as these allegations are mutually exclusive, only one can be correct; and, to be perfectly frank, I am by no means persuaded by the claims for either of them!

In 1992, the Danish weekly magazine *Hjemmet* published an article in which musicologist Prof. Lambert Wilson, a former conductor of the Aberdeen Symphony Orchestra, claimed that he had constructed a 'sea serpent' swimming mask, constituting the famous head and neck configuration in 'the Surgeon's photo', with tiny eyeholes enabling him to see, which he had worn while swimming in the loch on a day in late summer 1934. According to his story, this is what the London surgeon, coincidentally sharing the professor's surname, had seen and photographed. However, all other accounts concerning the taking of this photo placed the date in question in April, not during the summer. In any case, the very concept of someone swimming around in the loch with the model of a Nessie perched on top of their head is one for which I personally would be most reluctant to stick my neck out (so to speak!).

The second allegation received widespread publicity during March 1994, when researchers David Martin and Alastair Boyd announced that, only shortly before he died, in November 1993, expert model-maker Christian Spurling had confessed to them that the photo's image was actually that of a 1-ft-tall 'head-and-neck' model - constructed by him from plastic wood in January 1934, then attached to a toy clockwork submarine. This was secretly photographed (and afterwards sunk) in the loch. Surgeon Wilson (apparently fond of practical jokes, and supplied with full details concerning the model) was then recruited to publicise the photos. Martin and Boyd subsequently documented their investigations that had led them to put forward this startling claim in an absorbing book, *Nessie: The Surgeon's Photo Exposed* (1999).

Of especial note at this point is that a long-overlooked article appearing in London's *Sunday Telegraph* newspaper back on 7 December 1975 and written by 'Mandrake' (the pen-name of one Philip Purser) had actually claimed that Spurling's stepbrother Ian Wetherell had admitted to 'Mandrake' that he, while in the company of his father Marmaduke ('Duke') Wetherell and an insurance broker friend called Maurice Chambers, had taken some photos of a small toy submarine bearing the 'head-and-neck' made of rubber tubing in an inlet of the loch before sinking it, with one of these photos becoming the Surgeon's photo. Remarkably, however, this potentially significant article signally failed to be mentioned in any Nessie-related publication for almost two decades, which is a notable mystery in itself. More about this article a little later. Meanwhile, both of the Wetherells (and Wilson too) were dead by the time that Spurling made his own alleged confession, so they could not be questioned about it following its public release.

Spurling claimed that he had been requested to produce this model by his stepfather Duke and stepbrother Ian (who, Spurling alleged, also recruited Wilson as the fourth member of their conspiracy) - in order to avenge the recent public humiliation suffered by Duke when he rashly identified as genuine some footprints found on the shores of Loch Ness that were soon afterwards exposed as crude fakes. They had been produced by someone using an ashtray - not an umbrella stand, as claimed in many other publications - made from the foot of a hippopotamus! (Bearing in mind, however, that the Wetherell family actually owned such an object, it is by no means unlikely that Duke himself was the footprint hoaxer, but had not anticipated the ridicule that would rebound upon him.) As far as the quartet of hoaxers were concerned, there could be no better way of seeking retribution than to discredit the monster itself.

Sadly, the media's coverage in 1994 and since then of this supposed hoax was largely uncritical (indeed, it is still widely if mistakenly assumed to have been conclusively proved). So it was left to some astute newspaper readers to point out its array of inconsistencies, which can be summarised as follows.

The type of clockwork toy submarine available at that time could not have supported such an unwieldy structure as a model of the head-and-neck without being in serious danger of overbalancing; the only way to counter this would have been to place ballast inside the submarine - which would have promptly sent it plunging down beneath the water! True, in fairly recent times an ITN film crew and a Japanese film crew have successfully floated what they considered to be replicas of Spurling's 'head-and-neck' submarine. But as no-one has ever seen the latter toy (always assuming of course that it did actually exist!), having to rely instead entirely upon Spurling's verbal description of it and a picture of a toy submarine shown to him by Martin and Boyd that he claimed to be similar to the one that he had used, such accomplishments by the two teams clearly are by no means as impressive as they might outwardly seem to be.

The photo shows the head-and-neck surrounded by ripples - not to be expected if the craft were moving at the time of being photographed.

The 'head-and-neck' looks to be rather more than 1 ft tall. Indeed, based upon a comparison of the length of adjacent wind waves (with the wavelength estimated from modern results on wind waves and contemporaneous weather information), oceanographers Prof. Paul LeBlond and Dr Michael Collins calculated in 1987 that the neck's height above the water level was 4 ft.

Also, how can the 'model submarine' identity explain the second 'Surgeon's photo', whose 'head-and-neck' image has a very different outline from that of the first photo? Two contrasting explanations have been proposed – either the entity photographed was indeed alive, and had moved its position and orientation between the taking of the two photos; or the second photo was not depicting the same entity as the first, famous one (and may not even have been taken at the same time and/or location). Unfortunately, there seems no way of determining which (if either) of these explanations is the correct one.

Moreover, there are various noteworthy discrepancies between Ian Wetherell's confession as contained in the 'Mandrake' article of December 1975 and Spurling's own confession, which do not appear to have been highlighted anywhere before. According to Wetherell, the hoaxers were himself, Duke, and Chambers, whereas according to Spurling they were himself, Wetherell, Duke, and Wilson. So what had happened to Chambers? In the article, Wetherell is quoted as saying: "I took about five shots with the Leica", but if so, where are the others? Did they not come out? Only the Surgeon's photo and the second photo are known. Also in the article, the head-and-neck is said by Wetherell to have been made from rubber tubing, whereas Spurling claimed that it was made from plastic wood. Wetherell also claimed that the toy submarine was "only a few inches high", whereas Spurling stated that it was 1 ft high. How can these discrepancies be explained – a succession of memory lapses, or the failure of two brothers to synchronise their stories?

In addition, I am always very wary of 'death bed' confessions. Why wait so long, especially as the 'Mandrake' article had already been published way back in 1975 anyway? Even the faking of one of the world's most mysterious photos is not a crime (and releasing a supposed confession to having done so certainly hadn't troubled Ian Wetherell). Indeed, if anything, the release of the 'truth' while the hoaxers still lived might well have guaranteed them instant fame and lucrative financial gain (or at least it might have done for Wetherell had the article containing it not instantly sunk into absolute oblivion for two decades, by which time he had died).

A final curiosity is that in their book, Martin and Boyd also put forward as 'proof' that the Surgeon's photo was faked a very odd claim made in letters written during November 1970 by a Major Egginton, one of Wilson's friends, to Nessie researcher-author Nicholas Witchell. Egginton alleged that Wilson had told him that this photo had been faked by a keen amateur photographer friend of his (presumably Ian Wetherell), who had first of all snapped a photo of the loch and then, once back home, had superimposed a LNM model (presumably Spurling's?) onto the photographic plate. Martin and Boyd have claimed that these letters, retained by Witchell, constitute the most important evidence against the Surgeon's photo that has ever been uncovered, but in reality it is only anecdotal. In addition, few photographs have ever received such in-depth scrutiny as the Surgeon's photo, so if it had indeed been created by superimposing one image upon another back in the 1930s, when photographic techniques were far less sophisticated than they are today, I feel sure that this would have been exposed by now.

Moreover, the most surprising element of this particular aspect of the case, yet one that, strangely, Martin and Boyd seem not to have realised, as they have apparently not focused upon it during their investigations, is that if Egginton's claim *is* true, then a crucial component of the much-vaunted 'confessions' of Wetherell and Spurling is false. This is because the Egginton and the Wetherell/Spurling testimonies do *not* support one another; on the contrary, they fundamentally contradict each another. After all, if the Surgeon's photo were indeed the result of a Spurling-constructed 'head-and-neck' model borne upon a toy submarine being photographed by Wetherell directly at the loch and then sunk there (the proposal made throughout their book by Martin and Boyd), then simple logic dictates that it couldn't have been created by Wetherell at home by superimposing the model upon the plate of an earlier-snapped photo of the loch.

My own personal opinion regarding this episode is that the real hoax was not the 'head-and-neck' submarine but rather the *claim* concerning the 'head-and-neck' submarine. In short, the craft never existed - only the story of it, invented (with careless variations by them in their respective tellings of it) by two sons seeking revenge upon Nessie for their father's humiliation (and in so doing conveniently ignoring the fact that if, as seems highly likely, he'd hoaxed the footprints anyway, he was the author of his own downfall).

After all, why go to all the trouble of constructing such a craft (one that, by its very nature, is highly improbable) and perpetrating a hoax with it (with all the attendant risks of being caught in the act), when all that you need to do is to release many years later (in 1975 and 1993 respectively) a superficially plausible *story* of a hoax - one that cannot ever be conclusively proved or disproved by anyone else afterwards, yet which will nonetheless cast for ever more a deep shadow of doubt on a photo that has become over the years one of the most famous pieces of evidence in support of the Loch Ness monster's existence? Such a scenario would thus achieve the desired effect to the maximum extent yet via the minimum of effort. Consequently, this is much more likely to be the one that actually did occur.

Yet even if the craft did exist, how very convenient that it sank (or was sunk) directly after being photographed – thereby ensuring that it could never be found. Once again, therefore, as there is absolutely no physical evidence to confirm its original existence (no photographs of it before it was placed on the loch are known, nor any preliminary sketches of it with head attached or of the head itself, nor even any jotted preparatory notes about how it might be produced), why should anyone believe that it ever did exist?

After all, the ultimate irony here is that because there is no physical evidence to confirm its existence, only anecdotal, the LNM is dismissed out of hand by sceptics; and yet we are expected to believe unquestioningly in a Nessie-head-and-necked toy submarine for which there is no physical evidence either, once again only anecdotal.

A hoaxed photograph, or a hoaxed hoax? With Wilson, the Wetherells and Spurling all long gone now, no-one will ever know for sure. Meanwhile, the media and numerous online websites largely accept entirely uncritically that this iconic photo was a hoax, thereby unscientifically destroying the credibility of the LNM's single most significant piece of evidence, which is precisely what Spurling had hoped for. All in all, therefore, an excellent result for Spurling but a sad result indeed for modern-day cryptozoology.

There is no doubt from reading their book that Martin and Boyd conducted sterling detective work in pursuit of answers concerning the Surgeon's photo, in particular bringing to belated widespread attention the long-forgotten 'Mandrake' article. Yet somehow I fear that it is Spurling and the Wetherells who were the ones that truly destroyed the credibility of the most iconic cryptozoological image of all time – and all without having to do anything more tangible than spin a couple of very fanciful, inconsistent yarns.

In 1960, an amazing piece of cinefilm was shot by aeronautical engineer Tim Dinsdale at Loch Ness. After spending five unsuccessful days seeking the monster, it was now the morning of 23 April - the sixth, and final, day of his long-planned search. Driving along the loch's Foyers Bay stretch of road, he suddenly spotted a mahogany-coloured hump-like object protruding up through the water, roughly 1,300 yards away and oval in shape, with a conspicuous dark blotch upon its left side. By now he was out of the car and standing by the loch, observing the object through binoculars - and that was when it began to move.

Dinsdale began to film the object at once, shooting about 4 minutes of black-and-white film with his tripod-mounted 16 mm Bolex cinecamera as what he now believed to be some form of immense living creature swam away towards the far shore, generating a distinctive v-shaped wake and submerging slowly - before suddenly changing direction, swimming parallel with the shore in a southerly progression but still submerging until it had almost vanished beneath the water. Following its painstaking analysis of this remarkable film, the Royal Air Force's Joint Air Reconnaissance Intelligence Centre (JARIC) announced that in its opinion the object in the film "probably is an animate object" - in other words, part of a living creature, rather than a submarine, or surface craft such as a rowing boat. If so, then the creature is certainly huge, because JARIC estimated the hump to be 12-16 ft long, a cross-section through it to be not less than 5 ft high and 6 ft wide, and a speed for it of 7-10 mph.

Although a major piece of evidence supporting the reality of an animal of truly monstrous proportions, the featureless nature of the hump filmed by Dinsdale cannot shed any light on its owner's taxonomic identity - an all-too-frequent problem with Nessie data. In recent years, moreover, sceptics have sought to cast doubt upon the validity of the JARIC analysis's conclusion, by speculating that it was a boat after all, and even that what appeared to be the image of a larger body portion staying submerged beneath the water surface was merely a shadow of the portion above the water surface. To my mind, however, this latter claim in particular seems spurious. Perhaps the best summary of the respective claims offered in recent times by the film's critics and supporters can be found in Gareth Williams's book *A Monstrous Commotion: The Mysteries of Loch Ness* (2015). The most pertinent excerpt from that summary reads as follows:

> In 1999, Richard Carter, Adrian Shine and Dick Raynor tried to recreate the sighting under similar lighting conditions, aided by a 16mm clockwork Bolex camera and a dark wooden 15-foot boat that followed the same zigzag course across the Loch. They believed that the result looked 'remarkably like' Dinsdale's 'Monster'. This is in line with some alternative interpretations of the film. Shine and others have suggested that the 'Monster' can be changed into a man in a dark-hulled boat simply by watching the film on television with the contrast turned up. Shine later superimposed magnified images of the object taken from 170 frames using an 'image-stacking' method which removes the grain in the film and random artefacts. According to Shine, this reveals a 15-foot boat with one and possibly two passengers - and a lighter spot on the prow, where a circular licence-number plate would have been carried during the 1960s. His conclusion has since been endorsed by commercial image processing experts, including some of the JARIC personnel who produced the original report in 1966.
>
> Others beg to differ. Angus, the Dinsdales' youngest son, has also seen a modern reanalysis of the original 1960 footage. No boat magically creeps out of the grain; instead, a dark 'shadow' is revealed beneath the surface, showing a body, tail and a distinctive diamond-shaped flipper.

By virtue of its objectivity, sonar evidence is considered by the scientific community to be much more persuasive than eyewitness accounts, and even photographic evidence - which, unless of exceptional quality, is open to all manner of interpretation (not to mention increasingly frequent claims of hoaxing too, nowadays, due to the sophisticated techniques of digital photo-manipulation now readily available). From the late 1960s onwards, many research teams - including those from

Birmingham University in England, the British Museum, the Academy of Applied Science, and the Anglo-American 'Operation Deepscan' of October 1987 - have obtained important, reproducible traces of solid, seemingly animate objects encountered in midwater or near the sides or bed of the loch. They were swift-moving, exhibited profound diving abilities, measured around 20 ft long, and were readily distinguishable from shoals of fish or inanimate objects such as vegetable mats.

These thereby provide new insights into Nessie biomechanics and spatial distribution underwater. Once again, however, they have offered few morphological clues with which to identify its species - except, that is, for a remarkable occurrence in August 1972, which, in my opinion, yielded by far the most important evidence currently obtained in support of the Loch Ness monster's reality as a huge water beast of a still-undiscovered species.

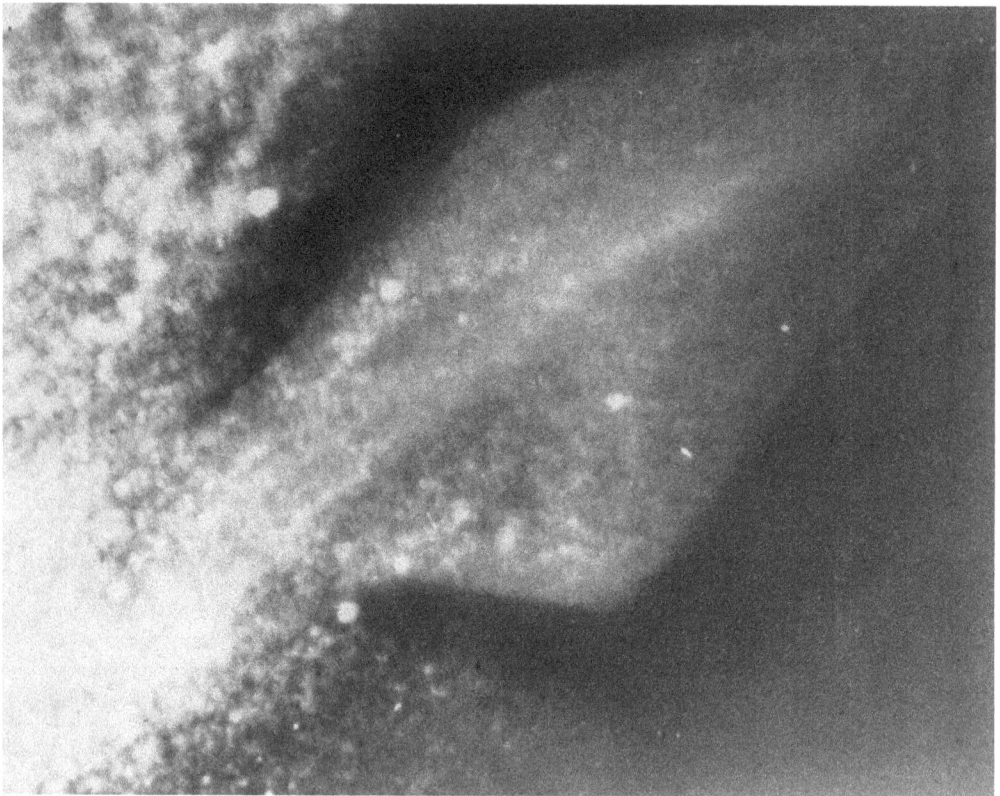

One of Dr Robert Rines's 'flipper' photographs
(© Dr Robert Rines/Academy of Applied Science)

With sonar apparatus and an underwater camera positioned in Urquhart Bay, Dr Robert Rines's team from his Academy of Applied Science were monitoring their equipment during the early

**Reconstruction of Nessie incorporating the diamond fins
of the 'flipper' photos (© Tim Morris)**

hours of 8 August when a flurry of movement was suddenly detected by the sonar. From the readings obtained, this appeared to be a shoal of fishes swimming rapidly away from something solid and much larger that was following closely behind - something that moved purposefully rather than passively drifting, and which measured 20-30 ft long.

Moreover, while the sonar had been obtaining traces of the body, the underwater camera had been photographing it. When its film was developed, two consecutive frames were found to depict something that computer-enhancement techniques revealed to be a diamond-shaped, pointed-tipped object resembling a hind flipper - whose orientation had changed slightly between the two photos, suggesting movement. Estimated to be 4-6 ft long, it was attached to a much larger body, and possessed what looked like a median keel or stiffening rod.

Intriguingly, it bore little resemblance to the flippers of pinnipeds, cetaceans, or other aquatic mammals, and did not correspond to the fins of eels or other ray-finned fishes either. Conversely, its outline strongly recalled the rhomboidal flippers of plesiosaurs.

A third, less-publicised photo on the 'flipper' film may portray two Nessies - depicting one lying on its side with two flippers visible, and the tail of a second close by. Correspondingly, the sonar traces showed that a second 20-30 ft object had indeed been present for a time within the vicinity of the sonar equipment and camera.

What makes these results so significant, and possibly unique at present, is that whereas eyewitness reports are subjective (and thus are open to criticism and doubts regarding their reliability and interpretation), here were two independently-obtained pieces of data (sonar traces and photographic film) that convincingly supported one another, and, of particular importance, were obtained by totally objective, disinterested witnesses - machines.

It was the distinctive image captured by the two 'flipper' photos that led on 11 December 1975 to the Loch Ness monster's formal scientific christening by Rines and British naturalist Sir Peter Scott - dubbing it *Nessiteras rhombopteryx* ('monster of Ness with diamond fin').

In addition, its shape corresponded perfectly with the dark, hairless, rubbery flipper that briefly broke the surface when, in September 1970, veteran American cryptozoologist Prof. Roy P. Mackal (until his retirement working in an official capacity as a biochemist at the University of Chicago) and two colleagues aboard a boat on Loch Ness spied a huge beast about 30 ft away, rolling underwater in Urquhart Bay.

(As with Dinsdale's film, however, there has been much in the way of sceptical discussion and claims regarding the sonar results and especially the 'flipper' photographs, and whether the latter may in fact owe their distinctive appearance to 'retouching' techniques applied to genuine photos of loch bottom sediment. Some of the sonar and photography-related issues are far too lengthy and technical to be presented here, but I recommend interested readers to consult various online sources, including http://www.lochnessinvestigation.com/flipper.html and http://blogs.scientificamerican.com/tetrapod-zoology/2013/07/10/photos-of-the-loch-ness-monster-revisited/ for further details.)

Nevertheless, there is still a heartfelt wish among Nessie believers for the creature to be spied in its entirety once in a while, rather than offering nothing more than fleeting glimpses of humps and long necks, or tantalising sonar traces of sizeable but anonymous bodies. In fact, such a wish may have occasionally been granted, because there are a few detailed eyewitness reports on file that claim sightings of Nessie as seen out of the water, moving around on land.

One of the earliest noteworthy instances occurred one afternoon in September 1919 on the shore of the marshland opposite Inchnacardoch Bay. A 15-year-old girl (the future Mrs Margaret Cameron), was playing on the bay's beach with her sister and two brothers when they heard a loud crackling noise coming from the marshland's trees - and then saw a huge beast emerge from among them. Pointing directly towards its amazed observers, it lurched down to the shore and into the water, humping its shoulders and twisting its small camel-like head from side to side upon a long neck. It seemed to be at least 20 ft long, with two short, round feet at the front of its body, a second pair further back, and a shiny grey skin.

During the 1930s, another group of children claimed to have seen a monster, this time in Urquhart Bay's bushy swamp. When later shown pictures of animals in a book, the creature selected by them as being similar to the one that they had seen was the plesiosaur.

The 1930s also yielded the Spicer sighting, one of the most dramatic Nessie reports ever recorded. On 22 July 1933, between 3.30 and 4.00 pm, Mr and Mrs George Spicer - visitors from London with no prior knowledge of Nessie - were driving along the road linking Dores to Foyers when they were confronted by a most incredible sight.

Emerging from the bushes about 200 yards further up the road, and raised several feet above its surface, was a horizontal object resembling an elephant's trunk - but it quickly resolved itself into a neck when followed swiftly by a thick, grey-coloured body estimated to be about 5 ft high. Its feet, if it had any, were not visible to its two eyewitnesses because the lower

Artistic reconstruction of Arthur Grant's land sighting of Nessie
(© William M. Rebsamen)

portion of its body was hidden in the slope of the road. Something protruded from the area of its shoulder, which the Spicers later deemed to be the tip of its tail, curling forwards along the side of its body facing away from them.

Its body moved via a series of jerks and its neck arched upwards as it progressed rapidly across the road and into the bracken separating it from the lochside. Judging from the width of the road, the beast's total length was put at over 25 ft, but moments later it had vanished, leaving no trace in the water. The Spicers were very repulsed by the beast's appearance: "It was horrible - an abomination...a loathsome sight" - forceful descriptions that one would not expect if its species was in any way familiar. Mr Spicer also referred to it as the "nearest approach to a dragon or prehistoric animal that I have ever seen in my life."

William McCulloch, a cyclist to whom the Spicers spoke soon after their sighting, lost no time in pedalling back to the spot where the creature had emerged, and he was able to confirm that the bushes on both sides of the road and leading down to the loch were extensively flattened, as if a steamroller had been driven over them.

Another memorable land sighting occurred 5.5 months later, and featured an exceptionally qualified eyewitness. At 1.30 am on the bright moonlit morning of 5 January 1934, 21-year-old veterinary student Arthur Grant was riding home on his motorbike, stone-cold sober and travelling from Inverness to Glen Urquhart, when he saw something dark move in the shadow of some bushes along the road's right-hand edge at the turn for Abriachan. What happened next is so startling that Grant's own account of it deserves to be quoted in full:

> I was almost on it when it turned what I thought was a small head on a long neck in my direction. The creature apparently took fright and made two great bounds across the road and then went faster down to the loch, which it entered with a huge splash. I jumped off my cycle and followed it but from the disturbance on the surface it had evidently made away before I reached the shore. I had a splendid view of the object. In fact, I almost struck it with my motorcycle. The body was very hefty. I distinctly saw two front flippers and there seemed to be two other flippers which were behind and which it used to spring from. The tail would be from 5 to 6 feet long and very powerful; the curious thing about it was that the end was rounded off - it did not come to a point. The total length of the animal would be 15 to 20 feet.

He gave further descriptive details at a meeting of Edinburgh's Veterinary Society:

> Knowing something of natural history I can say that I have never seen anything in my life like the animal I saw...It had a head rather like a snake or an eel, flat at the top, with a large oval eye, longish neck and somewhat longer tail. The body was much thicker towards the tail than was the front portion. In colour it was black or dark brown and had a skin rather like that of a whale. The head must have been about 6 ft from the ground as it crossed the road, the neck 3.5 to 4 feet long and the tail 5 or 6 feet long. Height from belly to the back would be about 4.5 feet and overall length 18 to 20 feet...

Inevitably, Grant attracted much ridicule. However, a search made shortly afterwards, led by a Fellow of the Zoological Society of Scotland called H.F. Hay, discovered some large flipper-like tracks measuring 24 in long and 38 in across amid shingle - some 70 yards further up the beach from where Grant had seen the beast go down the road's steep bank towards the loch. The team found some flipper tracks there too, spaced about 5 ft apart.

At 4.30 pm on 5 June 1934, a maid called Margaret Munro was looking out of the window of her employers' house overlooking Borlum Bay when she saw a very large beast about 300 yards away, turning back and forth mostly out of the water on the bay's shore. Watching it through binoculars for 25 minutes before it re-entered the water and swam away, she discerned a disproportionately small head perched upon the end of a giraffe-like neck, a huge dark-grey body with grey skin like an elephant but white underparts, and a pair of short front flippers. According to Munro, it was "...able to arch its back into large humps" - but whether the back physically arched itself, or whether it bore protuberances that could be raised or lowered to yield humps, is uncertain from this description.

A more recent land sighting occurred on the afternoon of 28 February 1960, when Torquil MacLeod spied an immense animal half ashore on the beach at Horseshoe, about 2.5 miles south of Invermoriston. With grey elephant-like skin and a long neck resembling an elephant's trunk, virtually throughout MacLeod's 9-minute-long, binocular-assisted sighting the creature faced directly away from him, but he spotted two paddle-like rear flippers - and once, as it turned around to re-enter the water, a front flipper also came briefly into view.

What makes this particular sighting so important is MacLeod's estimate of the beast's size - using the graticulations on his binoculars' lenses, he calculated that even without the tail, hidden in the water, the animal was 40-60 ft long! This is much bigger than anything reported by other eyewitnesses of terrestrial Nessies - twice as long, in fact. A plausible explanation of this anomaly, as put forward by Nicholas Witchell in his own coverage of these reports, is that perhaps it is the smaller, younger members of this species that come ashore most often - with the larger, fully mature members (like the MacLeod-observed specimen) spending most if not all of their time in the water. (Alternatively, MacLeod may simply have made an error in his calculations.)

After studying some of these accounts, even British zoologist and Nessie supporter-turned-sceptic Dr Maurice Burton was sufficiently impressed to suggest that it may actually be more rewarding to look for Nessie on land than in the water. Notwithstanding this, what remains unanswered is the question of why such a species should venture onto land in the first place.

Surely, creatures inhabiting a lake amply supplied with fish and other edible livestock scarcely need to seek prey out of the water. Reproduction would seem on first sight a more promising line of investigation - whether to lay eggs or to bear live young, many aquatic animals come ashore for this purpose. However, if Witchell's suggestion concerning the size of Nessie specimens generally reported on land is correct, then it would seem that these are juvenile individuals rather than adult, sexually-mature ones.

Yet even if the reason why such animals come ashore is still obscure, the apparent fact that they can move overland provides important support for eyewitness claims that similar animals inhabit a number of other Scottish lochs - including Morar, Shiel, Oich, Lochy, Quoich, and Arkaig (see Chapter 13). Perhaps this species actively migrates between lakes on a regular basis, or perhaps from one original inland home it has progressively invaded several others.

If we assume that Nessie does indeed exist, what could it be, zoologically-speaking? Numerous contenders have been put forward by LNM researchers down through the years, but as will be reviewed in Chapter 2, by far the most popular and persistent Nessie identity is an evolved, modern-day plesiosaur.

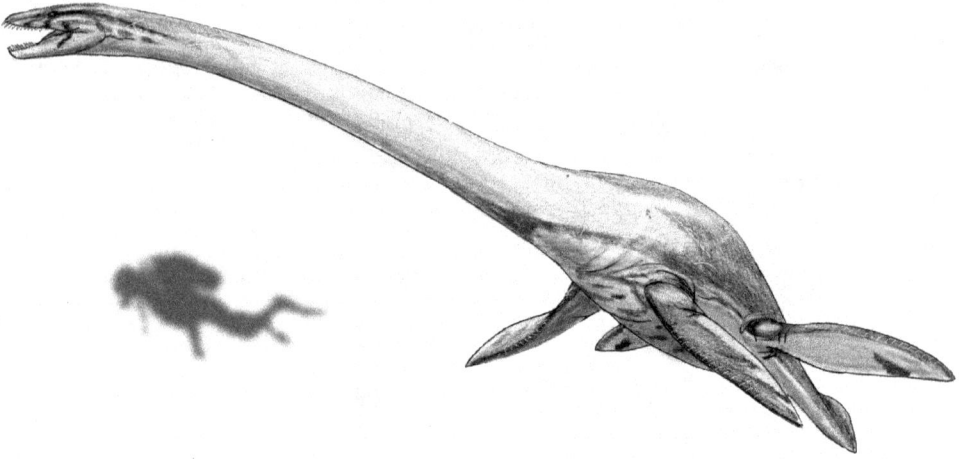

Modern reconstruction of the late Cretaceous North American plesiosaur *Elasmosaurus platyurus* (© Dmitry Bogdanov/Wikipedia CC BY-SA 4.0 licence http://creativecommons.org/licenses/by-sa/4.0/legalcode)

Modern reconstruction of the early Jurassic English plesiosaur *Plesiosaurus dolichodeirus* (© Dmitry Bogdanov/Wikipedia CC BY-SA 3.0 licence http://creativecommons.org/licenses/by-sa/3.0/legalcode)

CHAPTER 2:

REVIEWING THE LOCH NESS MONSTER
- PART 2: IN PRAISE OF PLESIOSAURS

Here and there high serpent heads projected out of the water, cutting swiftly through it with a little collar of foam in front, and a long swirling wake behind, rising and falling in graceful, swan-like undulations as they went. It was not until one of these creatures wriggled on to a sandbank within a few hundred yards of us, and exposed a barrel-shaped body and huge flippers behind the long serpent neck, that Challenger and Summerlee, who had joined us, broke out into their duet of wonder and admiration.

"Plesiosaurus! A fresh-water Plesiosaurus!" cried Summerlee. "That I should have lived to see such a sight! We are blessed, my dear Challenger, above all zoologists since the world began!"

Sir Arthur Conan Doyle – *The Lost World*

During the 19th Century, when considering the morphology of plesiosaurs, an English palaeontologist called Dean Conybeare likened them to "snakes threaded through the bodies of turtles". This is an apt description, for whereas their deep, inflexible bodies and two pairs of paddle-shaped flippers are reminiscent of those belonging to the great marine turtles, their very long, slender necks and small heads are certainly rather ophidian in superficial appearance - exemplified by the elasmosaurs. In these, the last known group of plesiosaurs, which apparently died out alongside the dinosaurs at the close of the Cretaceous, the neck was disproportionately long - in *Elasmosaurus* itself, it measured 26 ft in comparison with the 20-ft length of the entire remainder of the animal, and contained 71 vertebrae.

Plesiosaurs fed upon fish and water-dwelling invertebrates, and again like turtles they were almost exclusively aquatic, coming onto land only to lay their eggs. It is now believed that they swam like turtles too, moving their flippers up and down like great wings, so that they literally flew through the water (discussed in further detail later here).

In the light of this plesiosaurian précis, the selection of cases documented in Chapter 1 indicates that people are seeing creatures that look like plesiosaurs - but *are* they plesiosaurs?

Early plesiosaur illustration, from *Extinct Monsters - A Popular Account of Some of the Larger Forms of Ancient Animal Life*, 4th edit., by the Reverend H.N. Hutchinson, 1896 (public domain)

There is no doubt that the external correspondence between plesiosaur and long-neck is much more convincing than is true with most of the other cryptozoological identities on offer - giant eels, enormous worms or slug-like molluscs, specialised whales, huge sturgeons, or much-modified amphibians (this last-mentioned contender is rendered even less likely on account of the fact that hardly any marine amphibians are known, either living or prehistoric).

Nevertheless, there is a great deal more to consider than outward similarity when seeking to achieve a satisfactory zoological identification of a mystery beast. This concept was encapsulated by marine biologist Dr Forrest G. Wood from the U.S. Naval Ocean Systems Center, when commenting upon the nature of the object portrayed in Sandra Mansi's 'Champ' photo (snapped in 1977 at Lake Champlain straddling the Canada-USA border, Champ being North America's most famous lake monster) during a press conference held by the International Society of Cryptozoology on 22 October 1982 at the University of British Columbia:

> I will tentatively accept that it was a living animal. In appearance, it most closely resembles a member of the long-extinct group known as the plesiosaurs. That does not mean it is a

plesiosaur. All I can say is that, in general appearance, it most closely resembles a plesiosaur, which was an aquatic reptile. I accept it was a living animal, but I can't say what kind.

Some of the issues facing anyone attempting to champion the plesiosaur's candidature as the long-neck can be effectively dealt with by considering the classic case of the modern-day coelacanth *Latimeria chalumnae* - whose discovery in 1938 (followed by that of a second, closely-related species, *L. menadoensis*, in 1997) sensationally resurrected from extinction an entire lineage of prehistoric lobe-finned fishes, the crossopterygians.

PLESIOSAURS AND THE MISSING POST-CRETACEOUS FOSSILS
A major obstacle to the likelihood of plesiosaur persistence according to cryptozoological sceptics is the absence of verified post-Cretaceous plesiosaur fossils. Surely, they argue, if plesiosaurs had survived to the present day, we would have unearthed fossilised remains of them that bridge the gap between modern times and the close of the Cretaceous 66 million years ago. However, only quite recently has anyone identified any post-Cretaceous fossils of coelacanths either - yet *Latimeria* is conclusive proof that its lineage has indeed survived. There are also many other modern-day animals (so-called Lazarus taxa) with little or no fossil history linking them to prehistoric times (i.e. ghost lineages).

Interestingly, in 1993 American palaeontologists Drs Spencer Lucas and Robert Reynolds disclosed that since 1980 two samples of elasmosaur fossils (one a single cervical vertebra, the other a collection of about 40 incomplete cervical vertebrae) have been collected in California's Cajon Pass from a rock unit that may pertain to the San Francisquito Formation; some invertebrate fossils from this formation indicate a Palaeocene age. No direct association of known Palaeocene fossils and the plesiosaur remains, however, could be demonstrated. Hence Lucas and Reynolds decided to follow the more conservative option, and class the plesiosaur material as dating from the late Cretaceous. Yet a number of other supposed post-Cretaceous plesiosaur fossils have also been documented; these are traditionally discounted as so-called zombie taxa, i.e. deemed to have been reworked from earlier strata – but *have* they?

American cryptozoologist Scott Mardis has made a particular study of these controversial fossils, and a list summarising his findings made while searching through the palaeontological literature, presented in the 15 July 2014 post from Jay Cooney's excellent *Bizarre Zoology* blog, makes very interesting reading. Quoting from that list, here are its most intriguing entries:

1. Elasmosaurid plesiosaur fossils mixed with Paleocene microfossils in the Takatika Grit formation (Cretaceous-Paleocene) of the Chatham Islands of New Zealand.

2. The discovery of a specimen of *Plesiosaurus crassicostatus* in the Paleocene Waipara Greensand of North Canterbury, New Zealand.

3. The discovery of two sets of Elasmosaurid vertebrae (one articulated) allegedly associated with Paleocene microfossils in the Paleocene San Francisquito Formation near Cajon Pass, California. [See my above documentation of this case.]

4. A plesiosaur tooth in the Aruma Formation (Paleocene-Eocene) of Saudi Arabia.

5. A plesiosaur vertebrae [sic], assigned to the now-discarded genus *Discosaurus vetustus*, allegedly from the Eocene marine deposits of Choctaw Bluff, Clarke County, Alabama, deposits that have also produced specimens of basilosaurine whales.

6. A set of fossil vertebrae from alleged Cretaceous deposits in Mullica Hills, New Jersey is acquired by paleontologist Richard Harlan in 1824. He describes one of the vertebrae and assigns it to the Plesiosauria that same year. In 1851, paleontologist Joseph Leidy mysteriously reassigns Harlan's plesiosaur vertebra to the dolphin genus *Priscodelphinus* and declares it to be from the Miocene epoch. What about the age and identity of Harlan's other vertebrae? Is the age and identity of the *Priscodelphinus* vertebra completely resolved?

The most startling putative post-Cretaceous plesiosaur fossil of all – or, if adopting the orthodox alternative explanation, the most dramatically reworked plesiosaur fossil of all – must surely be the isolated dorsal vertebra of an elasmosaur that was found in a Pleistocene erratic boulder (geschiebe) near Wisbar, northern Germany, in July 2008, and formally documented in 2011. According to lithological studies conducted upon this fossil, however, it can be correlated with the 'Köpinge' sandstone from the Ystad-Vomb area in southern Sweden (which dates from upper Lower Campanian to lower Upper Campanian within the late Cretaceous). Consequently, despite its Pleistocene setting it is officially deemed to be of late Cretaceous age.

Similarly, as revealed by Dr Brian Witzke from the Iowa Geological Survey in an *Iowa Geology* paper from 2001, and subsequently documented by Mardis, reworked and transported Cretaceous fossils, including plesiosaur bones, are sometimes found in the glacial tills and associated gravel deposits in Iowa, especially in the western part of the state. Moreover, in an email sent to Mardis on 21 January 2011, Witzke revealed that the University of Iowa's paleontological collections contain one such specimen, undocumented in the literature at the time of the email, consisting of a plesiosaur metacarpal derived from glacial gravels on the south edge of Iowa City. These gravels are largely reworked from pre-Illinoian glacial deposits in the area (about 1 million years old).

Yet even if any of the above cases really do feature bona fide post-Cretaceous plesiosaur fossils, surely plesiosaurs could not have withstood the rapid diversification of aquatic mammals that occurred during the Cenozoic Era? Mardis feels that such a conclusion may not be warranted:

If we accept the possibility that the fragmentary plesiosaur material spanning from the Paleocene to the Pleistocene may not be "reworked", then there may be no 70 million year ghost lineage for plesiosaurs. Would not the emergence of marine mammals such as cetaceans and pinnipeds have severely hampered any potential post-Cretaceous comeback the plesiosaurs may have made, had a few squeaked through the K/Pg extinction event? Perhaps not. Plesiosaurs seem to have persisted through such ecological shifts as the presumed extinction of ichthyosaurs, the rise of mosasaurs and the coming and going of metriorhynchid crocodiles. Would the introduction of cetaceans and pinnipeds be that different? Some would argue that warm-blooded, highly intelligent marine mammals would have had a distinct advantage over any reptilian competitors. There is some evidence to suggest that plesiosaurs may have been homeothermic, lived in gregarious social groups and cared for their young.

Moreover, biologist Dr Michael Woodley, author of *In the Wake of Bernard Heuvelmans* (2008), a highly-regarded, in-depth assessment of Heuvelmans' classic but nowadays increasingly faulty sea serpent classification, has pointed out that the plesiosaurs' 'benthic grazer' ecological niche was not occupied by mammals following the latter reptiles' supposed extinction at the end of the Cretaceous. Surely, therefore, a niche remains for long-necked marine creatures (e.g. plesiosaurs) today?

The paradox inherent in the reworked plesiosaur fossils concept was masterfully summarised as follows by Jay Cooney in a *Bizarre Zoology* post of 9 May 2014:

> Looking at the litany of "reworked" plesiosaur fossils spanning from the Paleocene to the Pleistocene, I think the "no geologic evidence of plesiosaurs after the Cretaceous" argument is shot to hell. True, there is no unambiguous post-Cretaceous evidence that is not fragmentary. Nevertheless, there is some evidence. If only one isolated tooth, flipper bone or vertebra of a plesiosaur is found in a Mesozoic deposit, the paleontologists do not immediately invoke reworking to account for its presence in the strata where it was found. But let the same bone be found in a Cenozoic deposit, complete with other Cenozoic marine vertebrates, and it is immediately tossed into a refuse bucket labeled "reworked". Instead of marveling over this potential relict survivor, it is barely mentioned in the literature with disdain and then shoved into a drawer, hidden away to be forgotten or conveniently lost. If "reworking" will not get the job done, then one can always say it was mislabeled or even misidentified...There is a term for this. It's called "moving the goal posts". No doubt that the arch-nemesis of the Great Sea Serpent, Sir Richard Owen, would...approve.

Amen to that!

In view of the above considerations, it is clearly time that the above-listed and other ostensibly anachronistic plesiosaur fossils received proper radiometric dating (until recently, direct dating of vertebrate fossils had been unsuccessful), and also that the strata containing them were comprehensively, directly dated, in order to discover unequivocally whether such fossils have indeed simply been reworked from older strata or whether they are truly much younger than any previously confirmed examples.

A QUESTION OF HABITAT

Whereas fossil coelacanths were freshwater species, the modern-day *Latimeria*'s physiology is modified for a marine existence. Its evolutionary adaptation for survival in a profoundly different habitat from that of its ancient predecessors is of great relevance to the question of modern-day plesiosaur survival - because one of the arguments that has been frequently raised by sceptics in the past is that as the plesiosaur species currently known from fossils were marine, freshwater long-necks could not be plesiosaurs. Taking this line of thought even further: as the freshwater long-necks so greatly resemble marine long-necks that it seems reasonable to assume that they all belong to the same major taxonomic group of animals, if these freshwater versions are not plesiosaurs then surely their marine lookalikes cannot be either.

As pointed out as far back as the mid-1970s by Prof. Roy P. Mackal in his book *The Monsters of Loch Ness* (1976), however, some fossil plesiosaurs had been discovered under conditions that implied a freshwater environment, especially rivers and estuaries - and, as he also noted, we could speculate that while pursuing fish, or escaping predators of an inflexibly marine nature, these reptiles travelled up rivers and perhaps into lakes – a theory that would be famously espoused in the palaeontological literature by Dr Alfred Cruickshank during the 1990s. Propelling this hypothesis to its logical conclusion, it would not be unreasonable to imagine a situation whereby, finding ample food in large bodies of freshwater and freed from the problems of predation, some plesiosaurs might have gradually transformed into resident freshwater creatures - a scenario already recorded for such noteworthy forms as the seals of Lake Baikal and the sharks of Lake Nicaragua.

Is Nessie an evolved freshwater plesiosaur? (© Dr Karl Shuker)

And sure enough, in more recent years a number of smallish early Cretaceous leptocleidid plesiosaurs have been discovered that were indeed resident freshwater species, and were seemingly specialised for permanent existence within this specific habitat - rather than being there merely through retreating from marine environments into the relative safety of freshwater in order to avoid predation as postulated by Cruickshank. Of interest is that in 2012, a second plesiosaurian lineage, the pliosaurs, was also revealed to have a freshwater representative – *Hastanectes valdensis* – and other plesiosaurian groups are now thought to have evolved such forms too.

As the crocodilians never succeeded in invading many of the more temperate bodies of freshwater around the world, a vacant niche for a large reptilian predator in such habitats could conceivably be filled by plesiosaurs. And the seas are so unutterably vast that there would surely be more than sufficient space for a wide variety of aquatic mammals and reptiles to coexist without competing excessively with one another – as noted earlier, the niche for a long-necked benthic grazer filled by the plesiosaurs during the Mesozoic seems not to have been occupied by any known species during the Cenozoic.

FURRY ENDOTHERMIC PLESIOSAURS AND GIANT GIRAFFE-NECKED SEALS
Another purported problem to be faced when seeking to reconcile long-necks with plesiosaurs is that according to various eyewitness descriptions, the former sometimes possess features not exhibited by any of the fossil plesiosaurs on record. These include whiskers and other furry accoutrements, horns or snorkel-like projections, one or more dorsal humps, an absence of any discernible tail, and an unexpectedly flexible neck.

In his book *In the Wake of the Sea-Serpents* (1968), cryptozoologist Dr Bernard Heuvelmans

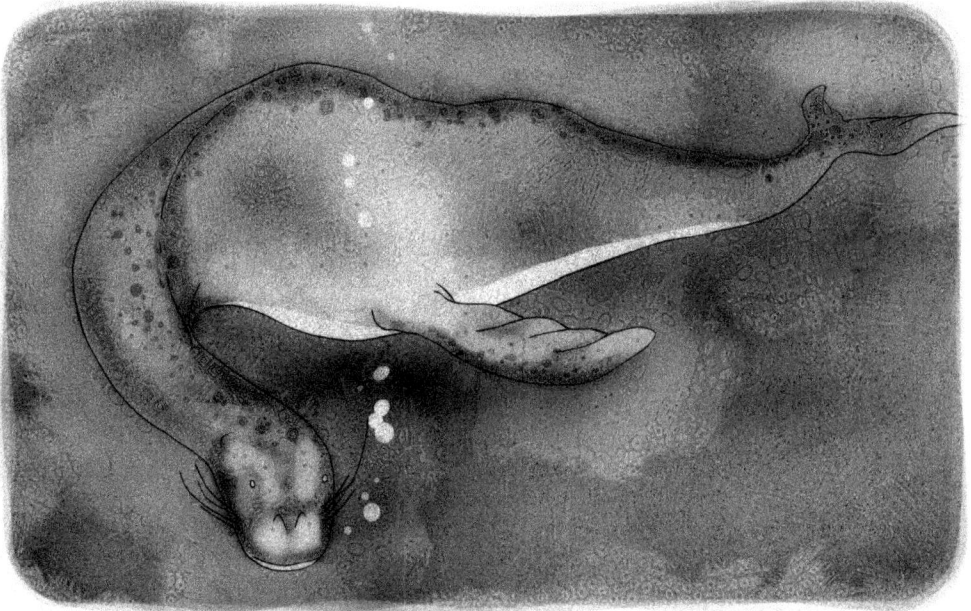

Nessie portrayed as a long-necked seal (© Robert Elsmore)

cited these discrepancies with traditionally-conceived plesiosaur morphology as good reasons for jettisoning the plesiosaur identity in favour of a novel mammalian contender - an undiscovered species of giant giraffe-necked seal, which he christened *Megalotaria longicollis*. He proposed that it occupies the ecological niche left vacant by the plesiosaurs' extinction, and thus, via convergent evolution, had acquired certain of their characteristics, most notably a lengthy neck. Peter Costello subsequently adopted this same identity for freshwater long-necks, within his book *In Search of Lake Monsters* (1974). (See Chapter 3 of this present book of mine for a comprehensive documentation and discussion of the long-necked seal's proposed candidature as the long-neck's identity.)

Although such a beast could be responsible for certain long-neck sightings, from the cases documented in Chapter 1 and from the discussion that follows here, however, it will become apparent that this identity cannot satisfactorily explain many others - especially those, for instance, in which a very distinct (and sometimes relatively long) tail is spied, which is not a typical pinniped feature.

In any case, in my view it is unnecessary to invent an entirely new animal (i.e. one that has no palaeontological predecessors) in order to explain long-necked water monsters when there are perfectly reasonable explanations for the morphological differences between such creatures and fossil plesiosaurs. After all, not even the so-called swan-necked seal *Acrophoca longirostris* from Peru and Chile's late Miocene to early Pliocene bears any realistic resemblance to the postulated appearance of Heuvelmans's hypothetical *Megalotaria*.

For instance, the 'hair' and 'whiskers' allegedly exhibited by some long-necks may not be true hair at all. It could constitute sensory filaments, composed of soft tissue that would not leave any impressions in fossilised specimens. Equally, the horns or snorkels might be protrusible breathing tubes. As for the dorsal humps: these could not be vertical undulations of the animals' bodies - unlike mammals, most reptiles cannot perform such movements (the long-extinct thalattosuchians or marine crocodilians were apparently a rare exception), and the framework of the plesiosaurian body was in any case extremely rigid. More reasonable are the popular suggestions that they are expanses of fatty tissue, inflatable airsacs, or even exposed portions of a dorsal crest - and, as with sensory bristles, it is unlikely that the outlines of breathing tubes, rolls of fat, airsacs, or membranous dorsal crests would be preserved in fossils. There is, however, another, much more important possibility - that these structures are, in evolutionary terms, a recent innovation.

All of what we know regarding plesiosaurs is based upon fossils that are at least 66 million years old, and which indicate that those ancient species were indeed unfurred. If, however, any plesiosaurs have survived to the present day, 66 million years of intervening, continuing evolution might have yielded animals markedly different from their Cretaceous ancestors - animals that may, for instance, have developed fur, and/or snorkels, airsacs, and crests (not to mention more flexible necks and loss of their tails - see later).

I have seen speculation regarding continued post-Cretaceous evolution for plesiosaurs and other officially long-extinct prehistoric creatures discounted as special pleading by some crypto-sceptics, but in my view it is exceedingly myopic to display this attitude, i.e. the wilful refusal to countenance that if a given taxon has survived beyond the date of its most recent fossils right through to the present day, millions of years of continued evolution are likely to have worked all manner of anatomical changes. This has happened with every major group of animals already known to be represented in today's fauna, so there are plenty of precedents for post-Cretaceous evolutionary change, including, once again, the coelacanths.

The living *Latimeria* is five times larger than many of its long-deceased fossil relatives, and, as noted earlier, is modified for a marine (rather than a freshwater) existence. True, during their Mesozoic reign, the plesiosaurs exhibited a relatively conservative degree of morphological transformation, at least macroscopically – but the changes that I have postulated above for an evolved, modern-day plesiosaur are not dramatically macroscopic either, as readily demonstrated by the fact of how outwardly similar long-necks seem to be to fossil plesiosaurs.

During the 1970s, a revolution occurred in dinosaur research - sparked by a dramatic theory promulgated by Dr Robert Bakker and given widespread public airing in Adrian Desmond's bestseller *The Hot-Blooded Dinosaurs* (1975). Namely, that at least some of the non-avian dinosaurs, hitherto looked upon as sluggish cold-blooded (ectothermic) creatures, might well have been active warm-blooded (endothermic) animals, just like mammals and birds.

Back in the late 1700s to mid-1800s, some naturalists deemed pterosaurs to be fur-covered bats (even marsupial bats, according to certain opinions!) rather than reptiles - a bizarre identification that was subsequently and discreetly discounted. However, following the remarkable discovery in Kazakhstan during the 1960s of a bona fide furry species of fossil pterodactyl, *Sordes pilosus* (whose 'fur' is actually composed of filaments termed pycnofibers that are not true hairs), it became increasingly likely that a physiology based upon warm-blooded principles (and irrefutably evident that the presence of a furry outer body covering) had indeed evolved independently in wholly unrelated groups of vertebrates. Since then, moreover, some researchers have proposed that certain therapsids (mammal-like reptiles) could also have been hairy; and it has of course been irrefutably confirmed by many palaeontological discoveries that certain dinosaurs, notably various theropods, possessed feathers, some of which were once again hair-like in form.

It is not impossible, therefore, that post-Cretaceous plesiosaurs (and perhaps even earlier ones) also evolved both of these traits, ultimately yielding endothermic furry species (see also the 2010 French study regarding endothermic plesiosaurs later in this chapter). In order to meet the high energy requirements for maintaining an endothermic metabolism, however, their habitat would need to contain an abundance of suitable food. Whereas this would not be a problem for marine plesiosaurs inhabiting the vast seas, freshwater versions might face problems if confined to relatively small, land-locked lakes - unless they were capable of travelling overland from one stretch of freshwater to another. For plesiosaurs frequenting giant biota-rich lakes like Champlain and Loch Ness, conversely, a small population could probably sustain itself without obligatory recourse to terrestrial migration.

Edward Neumann's famous engraving from 1843, portraying pterodactyls as furry marsupial bats! (public domain)

Leathery turtle engraving from 1896 (public domain)

In any case, also available for consideration is an alternative scenario that offers the benefits of endothermy without involving its endergonic burdens. One popular reason among cryptozoological sceptics for discounting the prospect of plesiosaur survival is that creatures resembling them have been reported in areas that are too cold to permit ectothermic reptiles to survive. Yet even if plesiosaurs never did acquire an endothermic metabolism, they could still exist in such localities - thanks to a physiological anomaly known as gigantothermy.

Ectothermic creatures weighing 1 ton or more lose body heat so slowly that to all intent and purposes they are functional endotherms (i.e. comparable to creatures that generate body heat metabolically, such as mammals and birds), and are thus termed gigantotherms. One marine vertebrate living today that has long and famously been thought to exhibit a gigantothermic physiology is the leathery turtle *Dermochelys coriacea*, which, with a verified maximum length of 7 ft, is far smaller than any plesiosaur potentially responsible for the long-neck reports on file.

Of course, young leathery turtles would be too tiny to be gigantothermic, but as they are born and mature in tropical zones this would not be a problem. Only as an adult does this species migrate into colder, northern waters, and once again the same could be true of plesiosaurs - though as they would exceed the minimum size threshhold at a younger age than the leathery turtle, they could begin their northerly migrations earlier.

Traditional image of marine plesiosaurs (public domain)

In recent times, however, further studies of the leathery turtle have revealed that its resting metabolism is not responsible for its body heat, so it is not gigantothermic. Instead, it derives its body heat from increased activity, constantly swimming but very rarely resting, thereby obtaining heat from muscle activity. Once again, however, the same principle might apply to young plesiosaurs of comparable size to adult leathery turtles.

In short, whereas the physiological mechanism of gigantothermy would satisfactorily facilitate the presence of adult plesiosaurs in low-temperature zones normally unfavourable to reptiles, it would not be sufficient in order to enable plesiosaurs to reside on a permanent basis in colder northern waters, such as Loch Ness or the Arctic seas, because this mechanism could not be utilised by young, small individuals. However, an increased-activity lifestyle comparable to that of the leathery turtle might enable them to exist there.

And indeed, in 2014 a research team that included palaeontologist Dr Matthew J. Vavrek from the Philip J. Currie Dinosaur Museum in Alberta, Canada, published a *Cretaceous Research* paper in which they documented the discovery in Arctic Canada of early Cretaceous non-marine plesiosaur fossils that were primarily juveniles but which nonetheless "would have been living in a region that experienced at least seasonally cool temperatures". In other words, young plesiosaurs were indeed able to survive in colder, northern freshwater localities after all. The source of these significant fossils was the Hauterivian–Aptian Isachsen Formation of Melville Island in Nunavut, Canada.

Moreover, there is a second way of keeping warm in such circumstances that could also have been developed by boreal plesiosaurs. Phocids or earless seals (i.e. not otariids or eared seals, which constitute the sea-lions and fur seals) swimming in Arctic or Antarctic waters contract the tiny blood vessels (arterioles) in their skin, which enables the skin to cool down to a temperature only marginally above that of the surrounding water. This prevents heat from leaving their body, so that whereas their skin is cold, the interior of their body remains warm - thereby effectively insulated from the low environmental temperature but without necessitating any metabolic rise in order to achieve this.

The leathery turtle also exhibits various anatomical specialisations for surviving in cold-water habitats. These include countercurrent heat-exchange mechanisms in their large flippers, fibrous fatty tissue and a thick skin saturated with oil, plus an insulative carapace.

Alternatively, the possibility of plesiosaurs evolving full, true endothermy during continued post-Cretaceous evolution has long stimulated speculation among palaeontologists and cryptozoologists alike. In 2010, however, a fascinating scientific study was published in the journal *Science* confirming that endothermy was in fact already present among Mesozoic plesiosaurs (as well as mosasaurs and ichthyosaurs), let alone any putative post-Cretaceous survivors. Including Dr Eric Buffetaut among the French research team responsible for this major breakthrough, here are the study's findings, as summarised in the paper's abstract:

> What the body temperature and thermoregulation processes of extinct vertebrates were are central questions for understanding their ecology and evolution. The thermophysiologic status of the great marine reptiles is still unknown, even though some studies have suggested that thermoregulation may have contributed to their exceptional evolutionary success as apex predators of Mesozoic aquatic ecosystems. We tested the thermal status of ichthyosaurs, plesiosaurs, and mosasaurs by comparing the oxygen isotope compositions of their tooth phosphate to those of coexisting fish. Data distribution reveals that these large marine reptiles were able to maintain a constant and high body temperature in oceanic environments ranging from tropical to cold temperate. Their estimated body temperatures, in the range from 35° ± 2°C to 39° ± 2°C, suggest high metabolic rates required for predation and fast swimming over large distances offshore.

In short, Mesozoic plesiosaurs were indeed able to maintain a higher body temperature than that of their external living environment. This in turn suggests a high metabolism adapted to predation and fast swimming over long distances, even in cold water. And to reduce body heat loss, plesiosaurs may even have developed blubber, just like whales did during the Cenozoic. So as plesiosaurs could live in cold, temperate locations after all, their alleged presence in Loch Ness, for instance, becomes less problematic from a purely physiological basis (always assuming of course that they originally made their way into it from the sea and became trapped there; geologically, this is a quite recent body of freshwater, only about 10,000 years old).

Moreover, in the case of putative evolved, modern-day plesiosaurs, the presence of blubber might explain the variable humps often reported with long-necks by eyewitnesses. Equally, their youngsters might not only be well-supplied with insulating layers of fat but also be profusely furred (like seal pups are). Although this is of course all entirely speculative, it offers up one very intriguing concept for consideration.

Sightings of otters are regularly reported from Loch Ness, and these lithe furry beasts have often been put forward as an identity for the Ness monsters - despite the fact that, if eyewitness reports and sonar traces are accurate, these latter beasts are very much larger than any form of otter. Perhaps in reality the exact *reverse* of this scenario is the truth - wouldn't it be supremely ironic if some supposed otter sightings actually featured juvenile hairy plesiosaurs??

THE TALE OF A PLESIOSAUR'S TAIL

In their respective books on sea serpents and lake monsters, Heuvelmans and Costello claimed that long-necks appear to lack tails, and that in sightings that seemingly featured tailed long-necks the eyewitnesses had mistaken the paired rear flippers of such creatures for tails. In reality, however, long-necks with two pairs of flippers *plus* a well-delineated tail have been clearly observed on a number of occasions, in the sea and in lakes - which thereby points more towards a plesiosaur than towards a pinniped in the identity stakes.

The most likely explanation why many long-neck eyewitnesses do not report seeing a tail is quite simply that the tail is not rendered visible when swimming. In whales and other cetaceans, fishes, and the extinct ichthyosaurs and mosasaurs, the tail is the principal locomotory organ - explaining why it is or would have been readily apparent to observers of such creatures. In the case of the plesiosaurs, conversely, the tail's function as a rudder would not offer any reason for it to emerge above the surface. Only during those rare occasions necessitating a plesiosaur to come ashore - to lay eggs or (if a freshwater species) to move from one lake to another one close by - would the tail acquire an additional role, assisting the neck in preventing the terrestrially-ungainly creature from overbalancing. (Traditionally, it was believed that plesiosaurs lacked a sternum or breast-bone - which would have made forays onto land very difficult - but in 1991, zoologists Drs Elizabeth Nicholls and Anthony Russell presented new evidence to challenge this assumption.)

In any case, even if the tail did become visible during swimming it might not be recognised for what it was. Heuvelmans noted that in certain cases featuring sea serpents of types other than the long-necks, tails of enormous lengths are described - which led him to suggest that these tails might actually have been trails of bubbles. If true, I see no reason why the reverse explanation could not apply in cases featuring allegedly tailless long-necks - i.e. that eyewitnesses have wrongly identified a relatively nondescript, inactive tail as a bubble trail generated by the swimming animal.

The concept of evolved plesiosaurs can also offer a clue in this tale of the long-neck's tail (or lack of it). In the elasmosaurs, the last plesiosaurs known from the fossil record, the tail was extremely short - and as propulsion was accomplished exclusively via their flippers, its function was limited to that of a relatively dispensable rudder. Thus it would not be surprising to find a further 66 million years of evolution dispensing entirely with this appendage, yielding tailless plesiosaurs. Consequently, even if some long-necks really are tailless, this still does not eliminate a plesiosaur identity from contention.

Another vibrant illustration of a plesiosaurian Nessie inspired by the 'flipper' photographs (© Peter Loh)

PLESIOSAUR FLIPPERS AND PLESIOSAUR FLIGHT (UNDERWATER!)

When our knowledge of a creature is based solely upon fossilised remains, our perceptions of its appearance, movements, behaviour, and general lifestyle can change quite drastically as new fossil discoveries are made and research upon them is carried out.

It was traditionally assumed, for instance, that plesiosaurs swam using their flippers as oars, i.e. moving them backwards and forwards in a horizontal plane, with the action of drag upon them providing the propulsive thrust. However, from studies of the shape of their pectoral and pelvic girdles, and the presence of excellent areas for muscle attachment on the ventral portion of these girdles enabling these animals' flippers to achieve powerful downstrokes, researchers now believe that plesiosaurs swam by underwater flying. That is to say, they moved their flippers up and down like penguins, sea-lions, and marine turtles, with propulsive thrust being provided not by drag but by lift. This belief is also supported by the pointed tips of plesiosaur flippers - flippers used as oars gain no advantage from being pointed, but they do if they are being used as wings.

This fundamental change of opinion is of relevance to the question of whether long-necks are indeed living plesiosaurs, because underwater flying yields a much greater wake than rowing, and many long-neck eyewitnesses have remarked upon the great displacement of water occurring as the creature swam by. Once again, this would argue against a seal identity as these swim by rowing rather than flying. Furthermore, there is also an item of evidence regarding long-neck locomotion that would appear to eliminate sea-lions - an important matter, bearing in mind that, unlike seals, these creature do swim by underwater flying. Namely, the remarkable 'flipper' photos obtained at Loch Ness in August 1972 by Dr Robert Rines's scientific team, because when closely scrutinised they ostensibly reveal a number of key clues regarding the taxonomic identity of the animal in question.

During his analysis of these photos for Witchell's book *The Loch Ness Story* (1989), veteran Nessie researcher Dr Denys Tucker, formerly of the Natural History Museum's Ichthyology Department, pointed out the presence of what seem to be two ribs visible beneath the skin of the animal's flank, just in front of the base of the flipper (which, as noted in Chapter 1, is apparently a hind flipper). Their position indicates that these are dorsal ribs, but whereas those of mammals do not extend as far back as the hind limbs, they do in plesiosaurs and other reptiles, implying a reptilian identity for the flipper's owner.

As for the flipper itself, it is ideally structured for underwater flying - relatively broad, flattened, pointed at its tip, and thin, but also rigid, thereby presenting a large, inflexible area of resistance to the water when held face-on during the propulsive downstroke, and enhanced by a novel non-skeletal flap at the trailing edge (yielding this flipper's distinctive diamond shape) to prevent turbulence. And during the passive upstroke, this flipper would provide first-class lateral stability, by functioning as a hydrofoil.

The presence of the long ridge-like structure running down the central axis of the flipper has been highlighted by opponents of the plesiosaur identity, who note that no such skeletal component has ever been recorded from the flippers of fossil plesiosaurs. Indeed, the only known animal of any type whose flippers display a similar structure is the Australian lungfish *Neoceratodus forsteri*. In reality, however, there is no proof that this mystifying ridge *is* a skeletal component. On the contrary, it is far more likely to be a wholly external keel, a refinement for increasing the streamlining of the flipper by enhancing water flow across it.

All in all, the enigmatic Nessie flipper is precisely what one might expect from the limb of an aquatic reptile of evolved plesiosaurian persuasion.

WHY THE PLESIOSAUR IDENTITY WINS BY A NECK

Views regarding flipper locomotion are not the only ones with cryptozoological pertinence to have changed over the years in relation to plesiosaur biomechanics. When the first plesiosaur remains were excavated and studied in Victorian times, scientific reconstructions of the living animals depicted them swimming at the surface of the ocean with gracefully curving necks held vertically erect, like surrealistic reptilian swans.

Created by sculptor Benjamin Waterhouse Hawkins (1807-1894), two of the
famous Victorian plesiosaur statues at the Crystal Palace Park in Bromley, London,
in which they are portrayed with highly flexible necks (© Dr Karl Shuker)

If an evolved modern-day plesiosaur could hold its neck vertically upwards out of the water as portrayed by this model, it would certainly be compatible morphologically with the iconic 'periscope' head-and-neck sightings reported for Nessie and other long-neck water monsters – but could it? (© Dr Karl Shuker)

Later, however, that image was replaced by one which portrayed plesiosaurs with quite inflexible necks that could not be raised or even curved to any great degree - a viewpoint succinctly expressed in 1914 by Chicago University researcher Prof. Samuel Williston:

> Textbook illustrations of the plesiosaurs usually depict the necks, like those of the swans, freely curved...But the plesiosaurs did not and could not use the neck in such a way. They swam with the neck and head, however long, directed in front, and freedom of movement was restricted almost wholly to the anterior part. The posterior part of the neck was thick and heavy, and could not have been moved upward or downward to any considerable extent and not very much laterally. From all of which it seems evident that the plesiosaurs caught their prey by downward and lateral motions of their neck.

This state of affairs persisted for many years, and for cryptozoology the inevitable outcome was that plesiosaurs could no longer be looked upon as serious contenders for the long-neck identity - because they were deemed incapable of swimming with their head and neck held high above the surface of sea or lake in the distinctive manner characterising this particular category of water monster.

That in turn enhanced the credibility of the plesiosaur's greatest challenger in the identity stakes, the hypothetical 'giraffe seal' *Megalotaria* (see Chapter 3 for a comprehensive coverage), because pinnipeds can certainly raise their head and neck up above the water surface. Nowadays, however, events have come full circle.

Based upon further studies and specimens, palaeontologists believe that plesiosaur necks were considerably flexible after all. For example, according to an authoritative encyclopedia of prehistoric life from 1988, with Profs. Brian Gardiner, Barry Cox, Robert Savage, and Dr Colin Harrison as consultants:

> The plesiosaurioids [i.e. the true, non-pliosaurian plesiosaurs] fed on modest-sized fishes and squid. Their long necks enabled them to raise their head high above the surface of the sea and scan the waves in search of traces of their prey.

Similarly, in the words of biophysicist Prof. R. McNeill Alexander, writing in 1989:

> If plesiosaurs had eaten worms or clams, we might suppose that they used their necks to reach down to the bottom, dabbling like ducks or swans, but their spiky teeth seem more suitable for catching fishes and squid-like animals which would probably have been too active to be caught easily that way. It seems likely that they darted at prey, extending their long necks to catch things as they swam by. The movement could have been fast, if the neck was held out of water. Herons use their long necks to dart at fish, though they stand in the water instead of floating as plesiosaurs presumably did.

In Bristol University palaeontologist Dr Michael Benton's *The Reign of the Reptiles* (1990):

> Elasmosaurs of the Cretaceous...took the long-necked adaptations to an extreme. Some had as many as 70 cervical vertebrae, and the neck could bend around upon itself two or three times. The elasmosaurs no doubt jabbed their snake-like necks rapidly among the scattering schools of teleost fishes. They could have darted the head in and out and seized several fish without moving the body at all.

Incidentally, although the presence of pronounced bony vertical processes (neural spines) upon the cervical vertebrae of plesiosaurs potentially decreased neck flexibility dorsally in these aquatic reptiles, the presence of cartilage and other soft tissue between the cervical vertebrae potentially increased such flexibility. Indeed, a *Comparative Biochemistry and Physiology* paper authored by M. Zammit and co-workers in 2008 claimed that certain elasmosaurs might have been able to raise their very lengthy necks dorsally by as much as 155°.

Also well worth bearing in mind in relation to the potential ability of plesiosaurs to poke their head and neck above the water surface is spy-hopping, in which the plesiosaur's body would be held vertically, not horizontally. So although the neck would project out of the water, it would in such instances still be in line with the body, not at right angles to the body, and therefore would not involve any dorsal flexion at all.

Early illustration of a plesiosaur, in traditionally-depicted swan-necked mode (public domain)

Moreover, as pointed out once again by Alexander in relation to plesiosaurs:

> If they swam under water with the long neck stretched out in front it would have been quite tricky for them to steer a straight course: if the animal accidentally veered slightly to one side, the water, striking the neck obliquely, would tend to make it veer more...It seems possible that plesiosaurs often avoided this problem by swimming at the surface with their necks out of the water.

Clearly, then, there is a sound biomechanical reason for plesiosaurs to be able to raise their lengthy necks above the water surface. However, long-necks certainly do not spend all of their time with their necks out of the water - otherwise they would have been seen much more often - so most of their swimming must occur while completely submerged. Yet in view of Alexander's remarks, surely such behaviour conflicts with a plesiosaur identity?

Not necessarily. Bearing in mind that even when long-necks 'up-periscope' through the water their necks are nowhere near as long as those of elasmosaurs and other overtly long-necked plesiosaurs, the simple conclusion would seem to be as follows. If plesiosaurs have indeed survived incognito into the present day, during their presumed 66 million years of post-Cretaceous evolution their necks have become shorter (at least by plesiosaur standards).

Biomechanically, a shorter neck would be much less of a hindrance to swimming while wholly submerged - thereby reducing any necessity for a plesiosaur to spend most of its time with its neck held out of the water. This in turn would enable plesiosaurs to propel themselves faster than those ancestors currently known from the fossil record, whose notably long necks - either when held out of the water, or when submerged under it - would have greatly impeded rapid swimming. The fact that some long-necks are said to be very fast swimmers has, until now, posed something of a problem for proponents of the plesiosaur identity, but it can be seen that an evolved plesiosaur with a moderately but not dramatically long neck could well be capable of a very respectable turn of speed, even with its neck submerged. Also, if they do generate heat by increased muscle activity as postulated earlier in this chapter, that would tie in very neatly with rapid swimming behaviour.

Moreover, an evolved plesiosaur with a more flexible neck, one whose vertebrae perhaps possessed less pronounced neural spines than those of fossil ancestors but more intervertebral cartilage and other soft tissue (all of which would increase neck flexibility), might conceivably achieve the 'periscope' head-and-neck appearances so frequently reported for long-necks by eyewitnesses. Equally, such a plesiosaur, with a moderate neck but nothing like the extreme lengths known for fossil elasmosaurs, would not face the problems of generating sufficient muscular power to lift the neck and maintaining its own balance in the water whilst doing so that an elasmosaur would have faced (and which therefore provides additional reasons why palaeontologists do not consider it likely that elasmosaurs could raise their necks to any significant extent vertically).

MONSTROUS BEHAVIOUR

Perhaps the classic sceptical response to the question of whether plesiosaurs have survived into modern times is the statement "If they had survived, science would have discovered them long ago". The informed cryptozoologist will soon point out, however, that this assumption is simply not warranted. As on so many other earlier occasions, the coelacanth *Latimeria* constitutes a notable precedent in favour of plesiosaur persistence - by remaining unknown to science until 1938. Indeed, as I revealed in my book *The Encyclopaedia of New and Rediscovered Animals* (2012), this is only one of many spectacular species of aquatic animal to have been discovered and described by science since 1900.

Among these are: several new species of beaked whale; a creamy-white river dolphin from China called the baiji *Lipotes vexillifer* (1916); the pa beuk or Mekong giant catfish *Pangasianodon gigas* (world's largest freshwater bony fish; 1930); the huge and monstrously grotesque megamouth shark *Megachasma pelagios* (1976); a bizarre sea-bottom worm called *Riftia pachyptila* that inhabits 8-ft-tall tubes and flourishes huge scarlet tentacles (1977); the Australian snubfin dolphin *Orcaella heinsohnri* (2005); and even, as noted earlier, a second coelacanth species, *L. menadoensis*, discovered off the Indonesian island of Sulawesi (1997). If all of these can evade scientific detection for so long, how can anyone claim with certainty that there are no reclusive plesiosaurs in existence?

If such beasts do exist, however, there must be reasons why they are so elusive - which an examination of long-neck behaviour as reported by eyewitnesses may help to uncover.

One of the principal arguments put forward by sceptics of the giraffe-necked seal identity against treating seriously its candidature as the long-necks' taxonomic identity is that if these latter beasts really were pinnipeds, then surely they would be seen far more frequently than is currently the case. This is because pinnipeds are exceedingly inquisitive and therefore very likely to make their presence readily apparent to anyone on or near the loch, and they also spend quite an amount of time ashore during the breeding season.

Aquatic reptiles, conversely, like crocodiles and water snakes, rarely show such interest in the presence of humans, and there is no reason to suspect that plesiosaurs would be any more inclined to do so. Moreover, if some plesiosaurs had become viviparous (bearing their young live), they need not even come ashore for reproduction.

Breathing air while staying unobserved would not pose any great problem for plesiosaurs either. Their nostrils' external openings are positioned very high on their heads (just in front of their eyes, in fact) rather than at the snout's tip. According to the traditional view concerning plesiosaur respiration, therefore, these beasts could lie directly beneath the water surface with only the tips of their nostrils protruding through it (and perhaps equipped with flaps or extrusible snorkels - the 'horns' noted by eyewitnesses?), thereby obtaining all the oxygen that they require while remaining undetected by all but the keenest-eyed observers.

In 1991, however, after studying a superb 200-million-year-old fossil skeleton of a 15-ft-long plesiosaur called *Rhomaleosaurus megacephalus*, Drs Arthur Cruickshank, Philip Small, and Mike Taylor revealed that it had possessed furrows that ran from the end of its snout and along its palate into each of its two internal nostril openings - which were positioned much further forward than those of such familiar aquatic animals as whales and crocodiles. This suggested that as the creature swam, water passing into the plesiosaur's mouth would travel along these furrows, into the nostrils' internal openings and out of them via their external openings. Consequently, the researchers speculated that these aquatic reptiles didn't use their nostrils for breathing, but rather for underwater sniffing - assisting them in their search for carrion and live prey.

How, therefore, did they breathe? Cruickshank, Small, and Taylor suggested that they accomplished this through their mouth - hence they would periodically come to the surface to gulp air. And sure enough, there are several eyewitness accounts of 'head-and-neck' sightings featuring Nessie and other long-necks in which the eyewitness has spied the creature making gulping sounds or movements, which would be entirely consistent with respiratory behaviour of this nature. Moreover, if these beasts chose merely to extrude the tips of their jaws above the surface, they could take in oxygen without being seen.

Plesiosaur behaviour also explains another initial anomaly of long-neck elusiveness - why are there no bona fide long-neck carcases on file? Today's advanced cetaceans are often stranded because of malfunctioning sonar and their inability to raise their necks out of the water in order to restore their sense of direction visually - whereas plesiosaurs, lacking sonar and potentially well-equipped with a flexible neck in evolved modern-day (and possibly also fossil) species, would not suffer such problems.

A scale drawing depicting the postulated size (and appearance) of Nessie in comparison to a human (© Connor Lachmanec)

There is, however, another facet of plesiosaur behaviour that is also of great pertinence to this issue. Like certain modern-day marine vertebrates such as penguins and sea-lions, plesiosaurs habitually swallowed large stones - unlike some cetaceans, seals, ichthyosaurs, and mosasaurs. For many years, zoologists were unable to suggest a reason for this behavioural dichotomy. The traditional explanation for stone swallowing was that they were utilised in grinding up food to facilitate digestion - but this is unsatisfactory, because, collectively, the members of one group ate much the same range of foods as those of the other group. Once again, it was Dr Michael Taylor, this time in 1993, who offered an alternative insight.

Taylor noticed that whereas the non-swallowers of stones all utilised their tail as the major organ of propulsion (the tailless seals use their hind flippers in an analogous fashion) and were fast swimmers, the swallowers relied principally upon their limbs, achieving a type of relatively leisurely 'underwater flight'. From this, he proposed that the swallowers were ingesting (and, if no longer required, regurgitating) stones for ballast purposes - which would counteract the tendency of such creatures to rise upwards through the water while swimming slowly or remaining stationary (the much faster speeds attained by some non-swallowers when hunting would prevent this from happening to them). More recently, however, the buoyancy theory for gastrolith function has been replaced in the opinion of some researchers by one favouring these stones preventing the swallowers rolling from side to side instead. But whichever theory is true, it would also mean that a carcase containing a fair number of gastroliths is more likely to sink down to the depths than to be washed ashore.

Of course, it could be argued that sea-lions also swallow stones but many sea-lion carcases are nonetheless washed up, so if long-necks are plesiosaurs, why don't their carcases do the same? Ironically, however, exactly the same argument can be levelled against the *Megalotaria* identity too, yet this is strangely overlooked by plesiosaur sceptics. In reality, known sea-lion species are primarily coastal, which explains why their carcases are sometimes found, whereas if long-necks are indeed plesiosaurs (or long-necked otariids) but stay for the most part in mid-ocean rather than coastal zones, the missing evidence of stranded long-neck remains may no longer be a mystery.

Although there do not appear to be any genuine long-neck carcases on file, there are certainly many records of supposed long-neck carcases, some decidedly plesiosaurian in appearance, but those that were formally investigated revealed a very different identity - a rotting shark. On 25 April 1977, for example, the corpse of a huge plesiosaur-like beast was caught in the nets of the Japanese fishing vessel *Zuiyo Maru*, trawling in waters about 30 miles east of Christchurch, New Zealand. About one month dead, the carcase measured roughly 33 ft long, and was in an extremely advanced state of decomposition, smelling so badly that the crew very speedily cast it overboard - but not before its measurements were recorded, a few fibres taken from it, and some photos obtained.

Biologists were very perplexed by the photos, and many identities were offered - including a giant sea-lion, a huge marine turtle, a whale, and a modern-day plesiosaur. For quite a time, its true identity remained undisclosed, until the fibres from it were meticulously analysed by Tokyo University biochemist Dr Shigeru Kimora. He discovered that they contained a very special type of protein, called elastoidin, which is found only in sharks - not for the first time, the scientific world had been fooled by a deceptively-shaped decomposing shark carcase. Indeed, this phenomenon is so well-known nowadays that cryptozoological writer Daniel Cohen has even coined an apt term for such a carcase - a pseudoplesiosaur.

Pseudoplesiosaurs generally turn out to be the decaying carcases of the basking shark *Cetorhinus maximus*, but as it looks nothing like a plesiosaur when alive, how do its dead remains acquire this appearance? In fact, the striking metamorphosis is easily explained.

When the carcase begins to decompose, the gill apparatus falls away, taking with it the shark's jaws, and leaving behind only its small cranium and its exposed backbone, which resemble a small head and long neck. The triangular dorsal fin also rots away, but sometimes leaves behind the rays, which can look a little like a mane - especially when the fish's skin also decays, allowing the underlying muscle fibres and connective tissue to break up into hair-like growth. Additionally, the end of the backbone only runs into the top fluke of the tail, which means that during decomposition the lower tail fluke falls off, leaving behind what looks like a long slender tail. The plesiosaur image is completed by the shark's fins, because the pectoral and sometimes the pelvic fins remain attached, yielding the plesiosaur's flippers.

Sometimes, however, the fins become distorted, so that they can (albeit with a little imagination!) look like legs with feet and toes, and male sharks have a pair of limb-like copulatory organs called claspers, which would yield a third pair of 'legs' - an extraordinary transformation from a dead basking shark to a hairy six-legged long-neck that explains the remarkable hexapod Stronsay sea serpent. A decidedly strange, giraffe-necked carcase, it was washed ashore in the Orkneys during October 1808, but its identity as a shark was confirmed by studying some of its preserved vertebrae, still held at the Royal Museum of Scotland.

Yet another likely reason why long-necks are not encountered as often as one might expect is that they dislike the sounds emitted by the various mechanical contrivances via which humankind puts to sea or traverses the larger inland waters of the world; hence they actively avoid them. Nessie, for instance, has apparently reacted instantly on several separate occasions to the sound of a motorboat engine, and in a number of cases the sounds of sea-going vessels seem to have generated sightings of rapidly-departing marine long-necks.

THE COURSE OF TRUE EXTINCTION NEVER DID RUN SMOOTH
So far, it has been shown that many objections commonly cited by sceptics for discounting the possibility of plesiosaur persistence into modern times are by no means as convincing as they seek to make out. However, there is one notable objection still to consider – the precise courses of supposed extinction taken by plesiosaurs and various other taxonomic groups.

Although coelacanth survival into the present day is commonly cited as a major precedent when considering prehistoric survival of other ancient lineages, and, as has already been discussed in various sections of this present book, it certainly has merit in this capacity, there is one area where it signally fails to provide the necessary degree of support.

Plesiosaurs and a number of other major groups of Mesozoic reptile (such as the ornithischian dinosaurs and non-avian saurischian dinosaurs, the pterosaurs, and the mosasaurs), as well as (albeit to a somewhat lesser extent) the ammonite molluscs, all exhibited high taxonomic diversity prior to a devastating collapse, an apparent mass extinction, by the end of the Mesozoic. In contrast, the coelacanths have always been far less taxonomically diverse and appear to have continued in this unspectacular but steady vein from the Mesozoic and onward through the Cenozoic into the present day, with no massive increase in diversity followed by a comparably massive collapse – or, at least, that is what the known fossil record indicates. But here is where the latter reveals its weakness as a basis upon which to make too substantial a judgement.

For although there are a fair number of Mesozoic coelacanth fossils on record, and two modern-day species alive today, the intervening 66 million years of coelacanth existence is represented entirely by the merest handful of remains, and which, in fact, constitute the only physical evidence preventing the entire coelacanth lineage from exhibiting a 66-million-year ghost lineage. So where are all of their missing fossils? Even if the coelacanths have been taxonomically sparse throughout their Cenozoic persistence, they should still be represented by more than the most paltry smattering of palaeontological proof for their occurrence during that era – unless, as noted earlier when considering the fossil record's shortcomings, such fossils do exist but only in terrain where they can never be unearthed (for physical, political, and/or financial reasons); or the coelacanths died in regions where fossils never or only rarely formed anyway. What if all of the above considerations are also true for Cenozoic plesiosaurs and other undiscovered Mesozoic survivors?

Certainly, there is little doubt that if plesiosaurs have indeed survived into the present day (and again this same argument can be applied to pterosaurs, mosasaurs, and non-avian dinosaurs too), they are severely depleted taxonomically and numerically, the plethora of ecological niches that they once occupied now overtaken for the most part by the explosion in mammalian and avian diversity that occurred during the early Cenozoic. However, certain specialised niches may still exist that they could occupy, albeit in reduced numbers and diminished diversity (in turn providing another reason why such creatures are not observed as often as one might otherwise expect). For instance, as noted earlier, Dr Michael Woodley has pointed to the 'benthic grazer' niche occupied by plesiosaurs during the Mesozoic seemingly remaining unoccupied by Cenozoic mammals. So although these long-necked reptiles' time as dominant alpha predators is undoubtedly long over, the possibility of contemporary plesiosaur survival in a more limited, specialised form is one that I find difficult to rule out entirely.

Finally: those readers who still consider a pinniped to be a more plausible long-neck than any plesiosaur might do well to recollect the Spicers' description of Nessie - "It was horrible - an abomination...a loathsome sight". Another eyewitness, Mrs H. Finley, exclaimed: "It was horrible! I wouldn't go to look at it if it was exhibited behind six-inch steel bars!". It is difficult to believe that any mammal, even one as extraordinary as a long-necked seal, would engender such profound feelings of revulsion.

Conversely, as Dr Denys Tucker so judiciously remarked, these are the very types of comment that we would not be at all surprised to hear from people visiting a zoo's reptile house. Just a coincidence?

Perhaps it is. Then again, as one Nessie researcher once drily observed: "If we regard everything as coincidence, we shall soon have to admit that Loch Ness has the highest 'coincidence rate' in the world!"

CHAPTER 3:

THE L⊕NG-NECKED SEAL
IN CRYPT⊕Z⊕⊕L⊕GY
- STICKING ITS NECK ⊕UT F⊕R NESSIE

> It is the Long-necked, which by convergence has acquired the outline of a plesiosaur, that has given credit to the notion that such reptiles survive – quite unjustifiably, since it is certainly a mammal.
>
> Bernard Heuvelmans – *In the Wake of the Sea-Serpents*

Although many mainstream palaeontologists may shudder at the merest thought of it, the Loch Ness monster's most readily-conceived public image will always be that of a typical plesiosaur – all neck, tail, and paddled limbs. Lurking in its shadow, never too far from scientific consciousness but a million miles away from popular recognition, however, is a second cervically-endowed yet very different identity candidate – the long-necked seal. Yet whereas the plesiosaur's at least erstwhile reality is unequivocally validated by the fossil record (albeit one in which this reptilian lineage is currently curtailed at a point over 60 million years ago), tangible evidence for the existence at *any* point in our planet's history of the kind of veritable giraffe-necked pinniped required to satisfy a mammalian identity for Nessie and other comparable 'periscope-profile' aquatic cryptids is conspicuous only by its absence. Indeed, to all intent and purpose there is no more proof for the reality of the long-necked seal than there is for the Loch Ness monster itself. So, when and how did this hypothetical horror come into theoretical being, and why does it persist in casting its nebulous shadow over the much more romantic (if no more realistic?) image of its plesiosaurian rival? It's time to find out!

A LONG-FORGOTTEN LONG-NECKED SEAL AT THE ROYAL SOCIETY
Although in modern times the concept of the long-necked seal as a zoological reality has been promoted most visibly by the cryptozoological triumvirate of Oudemans, Heuvelmans, and Costello, a mysterious creature not only fitting its description but actually referred to by that very same name had been documented as far back as the 1600s, but was completely

overlooked by cryptid chroniclers until the 1990s. This was when American cryptozoologist Scott Mardis made a highly significant discovery, by spotting its long-forgotten description on microfiche at the University of Vermont, after which he duly brought this surprising but potentially very important beast to present-day public attention at long last via an article published on 7 August 1996 in a Vermont weekly magazine entitled *Vox*.

In 1681, botanist Dr Nehemiah Grew published a catalogue of curiosities that could be found at that time in the museum of London's Royal Society. It was entitled *Musaeum Regalis Societatis: Or a Catalogue and Description of the Natural and Artificial Rarities Belonging to the Royal Society and Preserved at Gresham Colledge* [sic], and among the many specimen descriptions penned by Grew that it contained was one of a still-unidentified form of long-necked seal, based upon a preserved skin from an apparently young individual of this mystifying creature. Specifically referring to it as 'the long-necked seal', Grew described it on p. 95 of his catalogue as follows:

> THE LONG-NECK'D SEAL. I find him no where distinctly mention'd. He is much slenderer than either of the former [two other pinnipeds documented by him earlier – see below]. But that wherein he principally differs, is the length of his Neck. For from his Nose-end to his fore-Feet, and from thence to his Tail, are the same measure. As also in that instead of fore-Feet, he hath rather Finns [sic]; not having any Claws thereon, as have the other kinds.

Conversely, in most known species of pinniped the length of their neck is only about half the length of their lower body.

Grew's description was subsequently reiterated by James Parsons in a paper on marine seals published by *Philosophical Transactions*, a Royal Society journal, on 1 January 1751. In it, he listed various known species, and he included the long-necked seal within this list. Here is Parsons's slightly expanded version of Grew's original description of it:

> He is much slenderer than either of the former; but that, wherein he principally differs, is the length of his neck; for from his nose-end to his fore-feet, and from thence to his tail, are the same measure; as also in that, instead of his fore-feet, he hath rather fins; not having any claws thereon, as have the other kinds. The head and neck of this species are exactly like those of an otter...That before described [the long-necked seal], was 7 feet and an half in length; and, being very young, had scarce any teeth at all.

Accompanying its description, moreover, was an illustration of this unidentified creature (reproduced in Mardis's *Vox* article), which portrayed it with a decidedly elongate neck, and was captioned 'the long necked seal or sea-calf'. It was depicted alongside two other seals (the same two as described by Grew prior to the long-necked seal).

One of these two was termed 'the common seal' (i.e. *Phoca vitulina*), and was readily identifiable as this species. The other one, conversely, was more perplexing, being dubbed 'the tortoise-headed seal' (and which must wait for its own review elsewhere!). In his seal listing at the end of his paper, Parsons noted that the long-necked seal could be found "on the shores of divers[e] countries".

*Philos.Trans.Vol.*XLVII .TAB.VI.

The Royal Society's long-necked seal specimen as depicted in Parsons's 1751 paper (public domain)

Be that as it may, no additional skins of long-necked seals have been forthcoming since the time of Grew and Parsons – their specimen thus being unique. So where is this zoologically-priceless skin today – what may well be the only physical evidence of a cryptozoological long-necked seal ever obtained by science? Tragically, no-one knows – like so many other remarkable specimens of mysterious, unidentified creatures, it has seemingly been lost, vanished into that great void where cryptid material seems irresistibly and inexorably drawn, never to be seen again.

FROM OUDEMANS TO HEUVELMANS – AND FROM *MEGOPHIAS* TO *MEGALOTARIA*

Although, therefore, this was not its earliest appearance in the historical chronicles, the long-necked seal first made cryptozoological headlines during the early 1890s. This was when Dutch zoologist and passionate sea serpent investigator Dr Anthonie C. Oudemans envisaged just such a beast as the answer to one of the greatest riddles in 19th-Century natural history – the elusive identity of the even more elusive 'great sea serpent'.

After analysing numerous sea serpent reports originating from seas all around the world and dating back centuries in some cases, Oudemans considered that their most plausible explanation was the scientifically-undiscovered presence of an enormous species of pinniped, boasting a cosmopolitan distribution, and morphologically distinguished from all presently-known species not only by its huge size (capable of growing up to 200 ft long) and long slender tail (a very unseal-like feature), but, in particular, by its very sizeable, elongate neck (which bore a noticeable mane in the male). In illustrations depicting its likely appearance in life, it looked very like a mammalian plesiosaur (or a plesiosaurian mammal!).

Oudemans even gave this seagoing marvel its very own taxonomic binomial name – *Megophias megophias*. In so doing, he was classifying it as a new species within a (now-defunct) genus that had been coined back in 1817 by French-American naturalist and passionate sea serpent investigator Constantine Samuel Rafinesque-Schmaltz in his published description of an uncaptured snake-like marine cryptid responsible for a spate of sea serpent sightings reported at that time off Gloucester, in Massachusetts, USA (*Megophias* translates as 'big snake').

In 1892, Oudemans published his extensive study and conclusions in his now-classic tome *The Great Sea-Serpent*, which makes fascinating if frustrating reading. For at the risk of perpetuating further this unintentional bout of alliteration, his resolution of the sea serpent problem was fatally flawed. Anyone reading the vast array of sightings documented by him can readily perceive that the beasts observed belong to a variety of discernibly distinct types. Yet Oudemans, inexplicably, chose to shoe-horn them all into one, resulting in his creation of *M. megophias* as a 'one-size-fits-all' solution that was doomed to failure when attempting to convince mainstream scientists already highly suspicious of sea serpent reality that it was truly the taxonomic alter ego of this incognito maritime enigma.

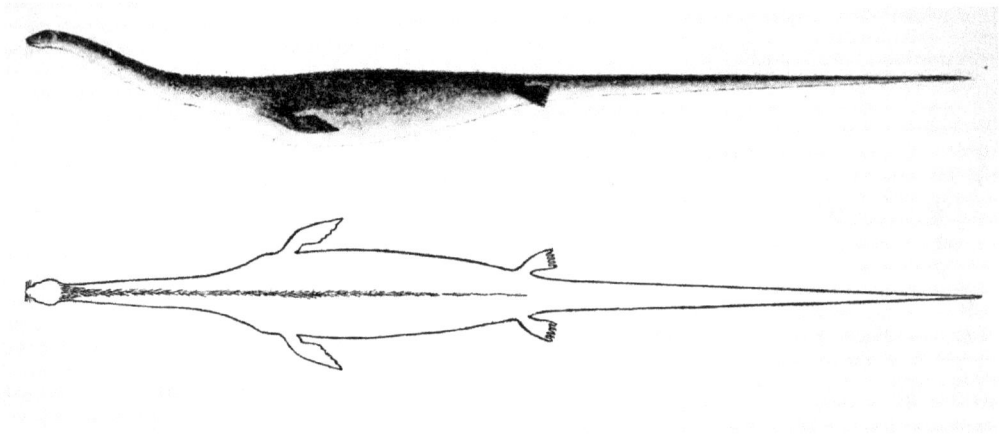

Oudemans's illustrations of his proposed long-necked (and long-tailed) mega-seal
Megophias megophias **(public domain)**

And so, inevitably, *M. megophias* floundered, Oudemans's ill-fated composite creation garnering little in the way of zoological credibility for itself, and rapidly sinking without trace into the gloomy abyss of scientific obscurity instead. And there it would linger, unloved and unlooked-for, all but forgotten for almost three-quarters of a century, until the long-necked seal hypothesis was finally retrieved, revived, and reconstituted in a very different form as part of a much more comprehensive, and complex, sea serpent classification conceived by a certain

Heuvelmans's nine postulated sea serpent categories (© Tim Morris)

Belgian cryptozoologist – Dr Bernard Heuvelmans.

Not only did Heuvelmans share a similar surname with Oudemans, when his own magnum opus on the sea serpent mystery was first published, in 1965 in French, it likewise shared the same name as Oudemans's – *Le Grand Serpent-de-Mer* ('The Great Sea-Serpent'). (Three years later, somewhat abridged and combined with a greatly-shortened version of an originally separate book on the giant squid and giant octopus, it was published in English as *In the Wake of the Sea-Serpents*.) And even his postulated long-necked mega-pinniped had a similar generic name to Oudemans's *Megophias* – namely, *Megalotaria*. But that is where the similarities ended.

Not making the same mistake as his near-namesake predecessor, in his grand scheme of sea serpent classification Heuvelmans conceived no less than nine distinct categories. Each constituted a different, scientifically-undiscovered species, and which (after discarding a sizeable number of accounts as unreliable or as outright hoaxes) he believed collectively explained all of the major sea serpent sightings reported from around the world down through history.

These hypothesised species were: a giant yellow tadpole-like creature of indeterminate taxonomic affinities; a gigantic 'super eel' (and/or a very elongate shark form); a marine reptile resembling a prehistoric mosasaur or a flippered crocodilian; an immense sea turtle; a many-humped serpentine zeuglodont-like cetacean; an armoured anomaly that he considered to be another zeuglodont due to his mistaken belief that armoured zeuglodonts were known from the fossil record (in reality, as later revealed, these were normal zeuglodonts whose remains had been found in association with armour-like scales derived from other, entirely unrelated fossil creatures); an exceedingly primitive stem cetacean of superficially otter-like form but much greater size and still possessing four limbs (his so-called 'super-otter'); and two separate types of pinniped, both of which were either tailless or near-tailless, like all modern-day species.

One of these pinnipeds, with a shorter neck, huge eyes, and a very noticeable mane, was dubbed by him the merhorse. The other, which combined the body and limbs of a typical otariid or eared seal (i.e. fur seals and sea-lions, possessing external ears) with an exceedingly long, giraffe-proportioned neck, he dubbed the long-necked (nowadays shortened to long-neck or longneck), and proposed for it the binomial name *Megalotaria longicollis* ('long-necked big otariid'). (In their 2003 book *The Field Guide to Lake Monsters, Sea Serpents, and Other Mystery Denizens of the Deep*, veteran American cryptozoologists/forteans Loren Coleman and Patrick Huyghe merged the merhorse and long-neck into a single sea serpent type, which they dubbed the waterhorse.)

After more than 70 years in zoological – and cryptozoological – exile, the long-necked seal was back!

From analysing 82 eyewitness accounts of alleged long-necked sea serpents, of which he deemed 48 to be certain, in his book Heuvelmans produced the following 'identikit' description of what he considered the likely morphology of this alleged cryptid to be:

A sea-animal of fairly large size, much bigger than the biggest pinnipeds and recognizable by its very long slender neck. Its general shape can vary greatly because of its thick layers of fat: sometimes cigar-shaped, sometimes serpentine when swimming fast, it may seem thick and stumpy when hunched up on itself. The relatively small head is round in shape with a somewhat tapering muzzle, sometimes like that of a seal or dog, sometimes like that of a horse, camel or giraffe. This apparent contradiction in testimony is doubtless due to the head lengthening with age, as is the rule among mammals.

The eyes are very small and can hardly be seen except from very close. In young ones there are a few whiskers on the muzzle. Two little horns can sometimes be seen on the head; these are probably erectile tubes arising round the nostrils. As the eyes are practically invisible, it is hard to place these tubes exactly in relation to them: at all events they rise from the top of the head. They would enable the animal to come to the surface to breathe without lifting its head out of the water, an arrangement like the skin-diver's schnorkel [sic]...

The neck is long and cylindrical; it is extremely flexible and can bend in any direction, especially in a vertical plane like a swan's. It may also stick perpendicularly out of the water like a telegraph pole. It has no mane, but a sort of collar, perhaps a fold in the skin, behind the head is sometimes mentioned.

The body is massive, thick and covered with rolls of fat so that it may, according as it bends, show one, two or three big dorsal humps, the middle of the three being the biggest. It has been suggested that these humps are inflatable air-sacs. This is possible, and the explanation cannot be excluded, but there is no need for any such theory in this case.

The spine forms a slight ridge all along its length, this may be due to a hairy crest or be accentuated by one.

There are four webbed feet, the front pair of which are often visible when the animal stands up vertically in the water, as the pinnipeds often do...When the hind feet are spread out in the same plane, they may sometimes look like a horizontal bilobate tail, as in the cetaceans. But they can also be held face to face, as the pinnipeds often do, and may then look like a fish's tail...

There does not seem to be much tail: at the very most it is a mere stump.

The skin looks smooth when it is wet and shining, but seen from close to it looks wrinkled and rough, like a walrus's or an elephant's. It is very dark brown on top, with black, grey or whitish mottling, while the underneath of the belly is dirty yellow and much lighter.

...Apart from one or two extravagant estimates of 200 feet or so, almost all the witnesses give a length between 15 and 65 feet – 60 feet often being given in round figures. There is, it is true, a series of witnesses who give lengths between 65 and 100 feet, and even as much as 120, but they seem to be influenced by the preconceived idea that it is a serpent, a plesiosaur, or even Oudemans's *Megophias*, and to assume it must have a tail as long as its neck and so extrapolate unjustifiably from the visible part of the body.

Speaking of extrapolating unjustifiably: I first read Heuvelmans's book over 30 years ago, and back then it seemed to me to be a work of superlative, near-genius zoological detection, worthy of the peerless if fictitious Sherlock Holmes himself (and indeed, Heuvelmans has actually been referred to as the Sherlock Holmes of zoology). In later years, conversely, as my own knowledge of cryptozoology, its methods, and its shortcomings increased, I re-read the book several times, and on each occasion with increasing scepticism regarding Heuvelmans's bold claims and intricate deductions.

Even taking into account the fact that he rejected many eyewitness reports as implausible, I personally feel that he nonetheless placed far too much emphasis upon the literal content of those that he did accept, i.e. he drew in-depth, often excessive, conclusions from the descriptions contained in those latter reports that, in my view, cannot be justified, because we simply have no idea just how accurate those descriptions really were.

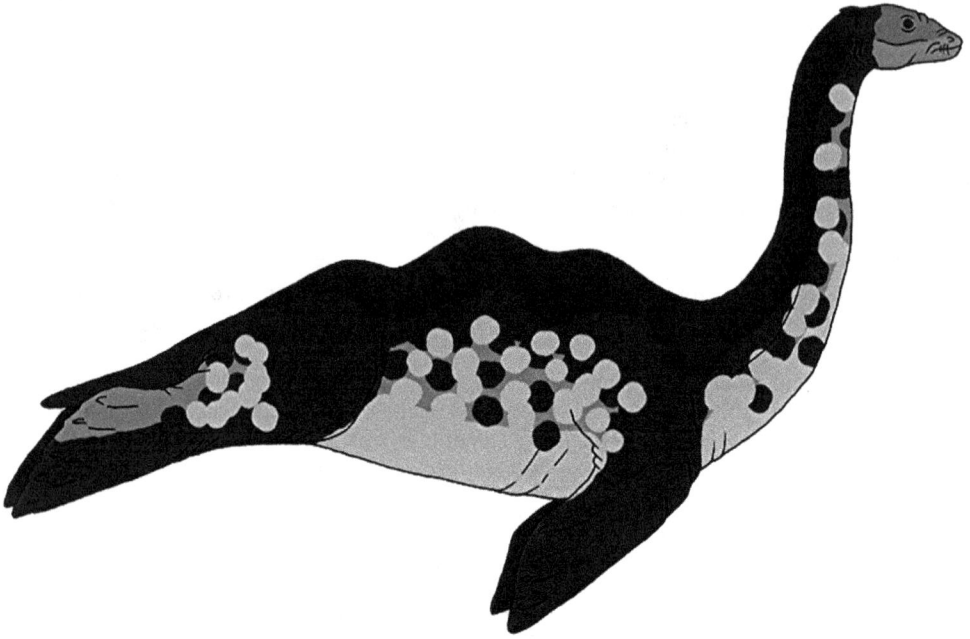

Representation of *Megalotaria* as Heuvelmans's long-neck sea serpent category (© Tim Morris)

From my own experiences of eyewitness accounts, I am well aware of the all-too-human failure of observers lacking a detailed knowledge of animals to describe with any notable degree of zoological accuracy the physical appearance of creatures that were unfamiliar to them (especially if doing so entirely from memory, and/or from a time some distance removed from the actual event and/or if they had encountered the creatures unexpectedly). This also applies to size estimates proffered by them. Such failure is surely responsible in no small way for the not-inconsiderable variations in eyewitness descriptions noted by Heuvelmans in his extensive and inordinately-detailed *Megalotaria* identikit account quoted by me above (and which is precisely *why* I reproduced it verbatim), most notably regarding the shape of its head, rather than (as he evidently if rashly believed) such variations being explicable entirely via anatomical or age-related phenomena.

In short, I believe that Heuvelmans placed far too much reliance upon the literal accuracy of eyewitness reports and far too little upon the likelihood that much of what was described in them were artefacts arising from poor zoological knowledge, inaccurate description, and flawed recollection.

In addition, following a close examination of Heuvelmans's sea serpent researches and his resulting nine-category classification system, German cryptozoologist Ulrich Magin argued in an extensive *Fortean Studies* paper from 1996 that far from being the outcome of an objective data analysis, Heuvelmans's sea serpent categories are subjective and predetermined, and that they don't actually function successfully when applied to individual cases. Magin's opinion is shared by British palaeontologist and cryptozoological author Dr Darren Naish, as expressed in a *Fortean Studies* paper of his own, from 2001.

Ideally, to avoid any subjectivity creeping into the data analysis when attempting to distinguish morphological categories of sea serpent present in the data, the analyser should be doing so blind, i.e. using eyewitness descriptions alone as the basis for creating sea serpent categories, not taking into account geographical localities or any other factors like Heuvelmans did. However, the analyser would then be vulnerable to falling foul of the uncertainty that invariably surrounds the accuracy of anecdotal evidence. For the most comprehensive examination and assessment of Heuvelmans's sea serpent classification, see Dr Michael A. Woodley's book *In the Wake of Bernard Heuvelmans* (2008).

All of the above criticisms also apply in relation to Heuvelmans's equally extensive, confident description of *Megalotaria*'s behaviour, yet once again based solely upon eyewitness testimony. According to his interpretation of such sources, this giraffe-necked maritime cryptid:

> ...is certainly the only sea-serpent that is amphibious. It is extremely flexible. The chief component of its movements is in the vertical plane; and this is mainly seen in its head swinging backwards and forwards when raised out of the water. This is also striking when the animal bounds on land, rhythmically gathering its hind legs up near its front ones and then leaping forward with the front ones, as the sea-lions do.

> Observers are often struck by the animal's staggering speed, which is quite exceptional at sea. Prodigious speeds, like that of an express train are mentioned, but more trustworthy witnesses, with more knowledge of the sea, generally give speeds between 15 and 35 knots. Such speeds seem to imply that it is a predator feeding on very fast-swimming fish. To catch its prey, the long-necked sea-serpent must make use of its long flexible neck to dart its jaws suddenly well ahead of its body.

> When the animal moves very fast turbulence waves appear on its very fat body as they sometimes do on the fatter pinnipeds, and this creates an illusion of small humps close together...

> No breath is ever visible. When the animal appears on the surface it sometimes leaves a greasy wake on the sea, as pinnipeds likewise do.

> A careful study of this type of animal...shows that its sight is rather poor...it must hunt its prey chiefly by sonar, as all the pinnipeds seem to do...

> It is evidently like a sort of huge gressigrade [i.e. otariid] pinniped with a very long neck, and more specialized than the sea-lions for a purely marine existence. It is true that this usually pelagic animal is still able to move on land, but it seems unlikely that it is obliged to go there to give birth: parturition must be able to take place at sea, a considerable advance over the sea-lions.

In addition to my above concerns regarding how literally he took eyewitness description, I also have some rather more specific criticisms of Heuvelmans's giant long-necked seal as the identity of the long-neck sea serpent. I have already aired some of these in Chapter 2, but there are a number of others too, as now revealed here.

For a creature as huge as *Megalotaria* yet only possessing tiny eyes, poor eyesight, and vibrissae present only in juveniles, utilising sonar for hunting its prey would not be an unreasonable prediction (as long as we remember that these above-cited characteristics are based entirely upon anecdotal, not physical, evidence). However, it is rendered far less plausible by the stark fact that even today, a full half-century after Heuvelmans wrote those above-quoted lines concerning this possibility, there is still no consensus that pinnipeds actually do employ sonar in hunting prey; over the years, this intriguing possibility has attracted many claims and counterclaims, but no conclusive evidence has been forthcoming. Nor has any for the possession of snorkel-like breathing tubes arising round the nostrils in any known pinniped species; so although such structures might indeed explain eyewitness reports of supposed horns, they would nevertheless be a notable evolutionary novelty.

The single most striking feature of *Megalotaria*, the one that earns for it its common name, is its exceedingly long neck. According to Heuvelmans, this neck "is extremely flexible and can bend in any direction, especially in a vertical plane like a swan's". One wonders, however, exactly how flexible did he mean by "extremely flexible", in view of the fact that as a mammal *Megalotaria* is exceedingly likely to have possessed only seven cervical vertebrae. It is the rotational and pivotal capacity of a vertical bony prong arising upwards from the axis (the second cervical vertebra) called the dens, which protrudes up through the ring-shaped atlas (the highly-specialised first cervical vertebra), yielding the atlanto-occipital joint, that enables the mammalian head (attached directly to the atlas) to turn through a considerable angle horizontally, and also to nod up and down.

But what about the rest of the neck? Assuming that it does contain only seven vertebrae, how feasible are Heuvelmans's claims about the extreme, swan-like flexibility of the neck of *Megalotaria*? One might expect from the vertebrae alone that through much of its length it would be as inflexible as a stiff rod, but neck flexibility in mammals is mediated to a considerable extent by the intervertebral cartilage discs and caps, so *Megalotaria*'s neck may well be more flexible than might otherwise be assumed. Also of note is that the giraffe's very long cervical vertebrae are connected to one another via ball-and-socket joints, thereby affording each section of the neck a remarkable degree of flexibility for a structure so exceptionally elongated yet composed of only a small number of very long internal prop-like structures. Might *Megalotaria* possess a comparable cervical arrangement? If so, however, this would be yet another major evolutionary novelty unparalleled among other pinnipeds.

Equally problematic is Heuvelmans's proposal that *Megalotaria* is a pelagic otariid that gives birth at sea, bearing in mind that otariids are in fact the most terrestrial of all pinnipeds, much more so than phocids or earless seals. For unlike phocids, the otariids can turn their hind limbs forward and are therefore able to walk on land. Also, they all breed on land, they come ashore more often than phocids (especially when moulting their fur), and often the adult males each maintain a harem of females on land (polygyny). Consequently, otariids are the least likely seals to have yielded a species exhibiting the predominately sea-living lifestyle that he envisaged for *Megalotaria*. Also, it is the phocids, not the otariids, that have also produced the biggest known modern-day pinnipeds – the two species of elephant seal *Mirounga* spp (which are even bigger than the walruses). And Heuvelmans's assertion that *Megalotaria* "bounds on land, rhythmically gathering its hind legs up near its front ones and then leaping forward with the front ones, as the sea-lions do", which does recall the terrestrial locomotion of otariids rather than phocids, was actually based upon just a single eyewitness account, so it is hardly a well-attested characteristic. Overall, therefore, it is more likely that if *Megalotaria* does exist, it is a phocid, not an otariid (*Megaphoca* as a replacement genus, anyone?).

Die dicknäsige Robbe. Phoca leonina. Lin.

Wonderful vintage painting of an elephant seal by Johann Bechstein, 1796 (public domain)

Heuvelmans discounted the possibility that the long-neck sea serpent sports anything but the shortest of tails – if, indeed, it possesses one at all. This clearly corresponds with a pinniped identity (a major problem with Oudemans's *Megophias* as any kind of seal was its very lengthy tail, because modern-day pinnipeds are conspicuously bereft of such a sizeable appendage). Yet he seemingly chose to ignore those eyewitness accounts that described long-necks with long tails. True, some such tails may have been artefacts, i.e. merely wakes or trails of bubbles, but others seemed genuine structures.

Heuvelmans concluded his coverage of *Megalotaria* by stating that apart from polar waters it exhibited a cosmopolitan distribution (an assertion drawn from the geographical distribution of eyewitness reports), generally sighted near the coast in cold temperate regions and in mid-ocean in warm temperate zones. Based upon more detailed analysis of the geographical spread of sightings plotted against the time of year when they have occurred, Heuvelmans further concluded that *Megalotaria* prefers spending the spring and warm season in northern cold temperate regions, migrating to the tropics to spend the end of the summer and the autumn there, before moving even further south into the southern hemisphere's temperate zone to spend the end of this latter hemisphere's summer there, thus avoiding entirely the cold extreme of the northern winter.

However, the reality of a highly mobile (and hence more readily encountered?) species of seal that is also "much bigger than the biggest pinnipeds" and occurs globally is one that I find difficult to accept. After all, the biggest pinniped currently known to exist today, the mighty southern elephant seal *Mirounga leonina*, already measures up to 22.5 ft long and can weigh over 10,000 lb – dimensions that are hardly inconsiderable.

Vintage photograph from 1936 depicting Goliath the elephant seal and his keeper at Vincennes Zoo, Paris (public domain)

2007
YEARBOOK

Edited by
Jonathan Downes

A long-necked seal drawn by Mark North, on the front cover of the *CFZ 2007 Yearbook*, which contains Cornes's very detailed article on this hypothetical pinniped (© Mark North/CFZ)

Yet if Heuvelmans is to be believed, this latter pinniped is positively dwarfed by a truly colossal species that is three times its size, and whose long-necked morphology sets it even further apart from all other pinnipeds, but which, incredibly, is still unrepresented by a single specimen. Not even so much as a beached skull or skeleton portion appears ever to have been discovered and retrieved on any coast anywhere in the world, despite Heuvelmans's assertion that the long-neck is of cosmopolitan distribution, and whereas occasional remains even of exceedingly little-known and quite possibly uncommon species of beaked whale and other very large, exclusively maritime mammals have indeed been found washed ashore.

In 2007, an extensive 117-page article written by cryptozoological enthusiast Robert Cornes that supported the possibility of the long-neck sea serpent and its freshwater counterpart constituting some form of undiscovered long-necked pinniped was published in that year's *CFZ Yearbook*, and included a number of thought-provoking speculations. One of these was that perhaps this surreal seal does come ashore to breed (rather than doing so at sea, as proposed by Heuvelmans) but remains unseen while on land by breeding in remote, inaccessible caves. Bearing in mind that seals breeding on land is generally not only a very visual affair but also a very noisy one, it would surely require a highly secluded location indeed for *Megalotaria* to breed while remaining out of earshot. Another speculation concerned whether the lengthy neck may assist in thermoregulation, in a manner reminiscent of one confirmed with seal flippers – in which warm blood can be diverted into these limbs, after which they are waved in the air to assist the animal in cooling off.

Also in 2007, sea serpent researcher Bruce A. Champagne published a comprehensive article entitled 'A classification system for large, unidentified marine animals based on the examinations of reported observations' within the multi-contributor tome *Elementum Bestia* (edited by American cryptozoologist Craig Heinselman). Like Heuvelmans, Champagne differentiated nine different sea serpent types, but they did not all correspond with Heuvelmans's; moreover, he also subdivided some of these types to yield several subtypes.

One of Champagne's nine types was the long-neck, which he then split into two subtypes, distinguished primarily via the size of the head in relation to the neck diameter. Most long-neck sightings were assigned by him to the first subtype, in which the head's diameter was the same as or slightly smaller than that of the neck. In addition, and going totally against Heuvelmans's opinion, Champagne proposed that this long-neck subtype sported a long tail (thereby hearkening back to Oudemans and *Megophias*). The second subtype, in which the head's diameter was larger than the neck's, consisted of five sightings from the North Atlantic off Great Britain and Denmark, and all five of these featured robust animals that, according to the eyewitnesses, were over 55 ft long, and therefore much bigger than those long-necks constituting the first subtype, which did not exceed 30 ft at most.

To me, the long-neck sea serpent is an enigma – a cryptid that I want so much to exist, as it would solve many cryptozoological riddles (and not just marine ones either, as I am now about to reveal), but which, at least in the guise of *Megalotaria* as envisaged by Heuvelmans, seems beset by serious shortcomings.

HEUVELMANS
LONG-NECKED

CHAMPAGNE
1A- LONG-NECKED

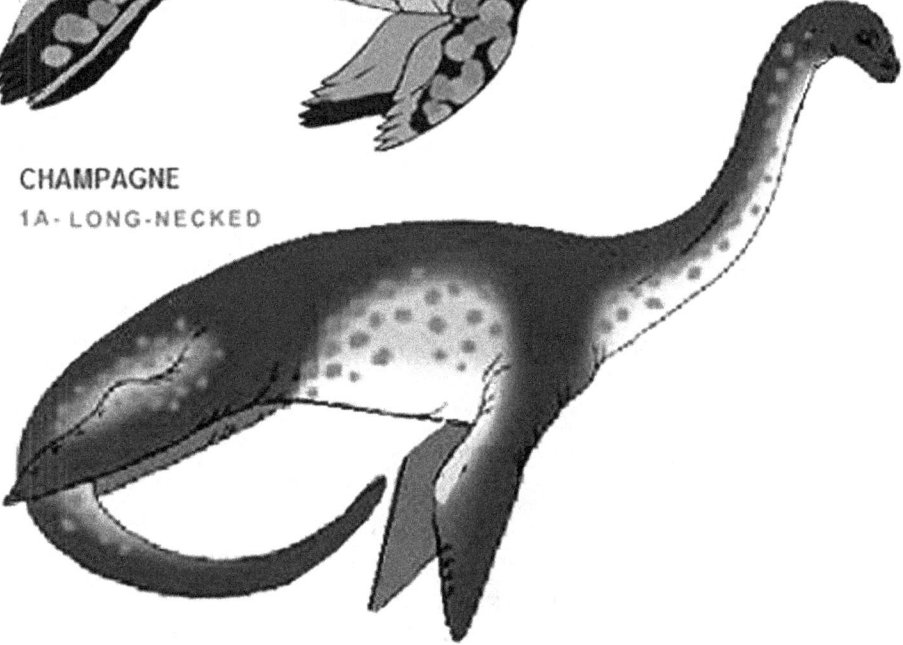

CHAMPAGNE
1B- LONG-NECKED

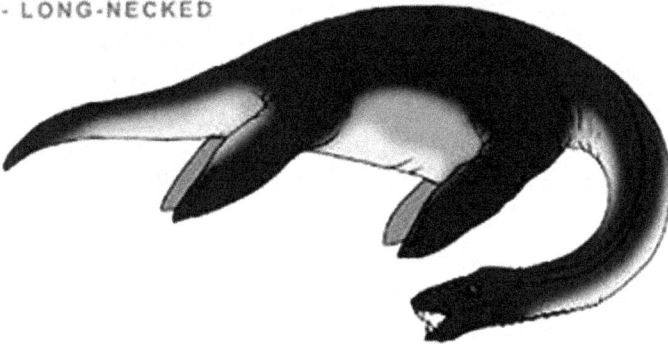

Champagne's two long-neck subtypes, compared with Heuvelmans's *Megalotaria* long-neck (© Tim Morris)

MEGALOTARIA, MEET NESSIE!

Heuvelmans believed that it was his hypothesised long-necked seal *Megalotaria* rather than any postulated form of evolved surviving plesiosaur that was responsible for those water monsters yielding the now-iconic, vertically-held, periscope-like head-and-neck image firmly planted in everyone's mind when picturing water monsters (and most especially the Loch Ness monster), whether marine or freshwater in habitat, though in his book he confined himself to those cryptids on record from the seas and oceans.

Just under a decade later, however, one of Heuvelmans's cryptozoological disciples and longstanding correspondents, Irish author Peter Costello, produced what was very much a companion book to his mentor's sea serpent tome but concentrating its attention instead upon lake monsters, in particular Nessie. (Judging from a footnote in his sea serpent tome – "Which will appear in a separate book on 'monsters' of lochs, lakes, marshes and rivers – freshwater unknown animals" – apparently Heuvelmans had originally planned to prepare such a book himself, but subsequently assisted Costello in producing his own book instead.)

If Nessie were a long-necked seal? (© Robert Elsmore)

Published in 1974, Costello's book was entitled *In Search of Lake Monsters*, and in this global study he followed much the same course as Heuvelmans did in his own, i.e. analysing an extensive collection of eyewitness reports of aquatic cryptids from around the world (but freshwater in this instance, with particular emphasis upon Scottish loch monsters), and then providing what he considered to be the most likely identification for them. Here, however, he diverged markedly from Heuvelmans, pursuing the Oudemans approach instead.

For whereas Heuvelmans had proffered a series of no less than nine different hypothetical cryptids as the collective solution to the sea serpent mystery, Costello bravely put forward only a single identity to explain virtually all of the lake monsters documented by him (including Nessie), diverse though they seemed to be in form, and therefore potentially inviting criticism of the kind that Oudemans's *Megophias* had attracted, i.e. that his solution was of the 'one-size-fits-all' variety – but that was not all. The single identity that he proposed was none other than Heuvelmans's very own giant long-necked seal, *Megalotaria longicollis*, thereby deeming it to be capable of living in freshwater habitats as well as in marine environments.

(Interestingly, although little-known, is that four decades after the publication of his sea serpent tome, Oudemans also examined the Loch Ness Monster's history, producing a short work entitled *The Loch Ness Animal*, published in 1934, in which, like Costello, he considered a single species to be responsible for both marine sea serpent sightings and freshwater Nessie sightings, thus believing that the latter were due merely to *Megophias* specimens entering the loch from the sea, but only on a temporary basis, not residing there permanently.)

As expected, therefore, for the most part Costello's description of this giant long-necked seal reiterated that of Heuvelmans for the same hypothetical species. However, he did also provide a few additional details, especially when specifically relevant to its inhabiting a freshwater domain - such as the assertion (rather than merely a speculation as offered by Heuvelmans for maritime *Megalotaria*) that it hunts by sonar, especially in peaty, stygian bodies of water like Loch Ness (in which underwater visibility is no more than 4 in), where vision is rendered largely or entirely superfluous; and that its hearing is therefore exceptionally sharp. As I noted earlier in this chapter, however, currently there is no conclusive evidence that pinnipeds do use sonar. He also claimed that it gives vent to a sharp staccato cry that sounds like a sea-lion's bark.

According to Costello, therefore, Nessie is merely a lake-dwelling long-necked seal, a freshwater-confined representative of Heuvelmans's marine *Megalotaria*, not even sufficiently distinct, despite its different habitat, to warrant any taxonomic delineation from the latter creature. Yet if this were true, why have other maritime pinnipeds only rarely or never established exclusively freshwater intraspecific populations? The only notable examples are two totally freshwater subspecies of the ringed seal *Pusa* (=*Phoca*) *hispida* – namely the greatly-endangered Saimaa seal *P. h. saimensis* (confined entirely to Finland's Lake Saimaa) and the Ladoga seal *P. h. ladogensis* (confined entirely to Russia's Lake Ladoga) – and some non-taxonomically discrete colonies of the common seal *Phoca vitulina* in a few lakes, such as Alaska's Lake Iliamna (already well-known to monster seekers for the giant fishes that allegedly inhabits its voluminous waters) and certain lakes in Quebec (a few researchers do elevate these Canadian individuals to the rank of a valid subspecies of common seal, known as the Ungava seal *P. v. mellonae*).

For the most part and with the vast majority of pinniped species (particularly the bigger ones), however, colonisation of freshwater simply does not occur. Yet it's not as if they never find their way inland from the sea – on the contrary, every year there are confirmed reports of seals in various rivers across the UK, for instance, and there are even verified records of specimens of known seal species in Loch Ness itself. However, whereas these have not led to the establishment of landlocked freshwater seal colonies (despite being much smaller than *Megalotaria* and therefore enabling a given volume and prey content of freshwater to accommodate and sustain more specimens of these seals than would be the case with a giant long-necked seal), according to the freshwater long-necked seal hypothesis the marine *Megalotaria* has somehow managed to accomplish this feat in numerous lakes all across the world.

But how could this particular pinniped species (always assuming that it does exist, of course!) have been so markedly successful at freshwater colonisation on an international scale, which would surely have involved some very visible migrations into freshwater at the onset, while also being so extraordinarily (indeed, inexplicably) adept at eluding all attempts by scientists and laymen alike to confirm its reality that not so much as a single skull or skeleton has ever come to scientific attention anywhere across its entire global distribution?

It is just about within the realms of possibility that amid the vastness of the world's seas and oceans the maritime *Megalotaria* can still evade scientific detection even in modern times, but how can its freshwater counterparts do the same, even when their lakes occur in close proximity to human habitation? For me the concept of *Megalotaria*, whether in the seas or (especially) in freshwater lakes, remains a particularly thorny one both to grasp and to retain.

SWAN-NECKED SEALS IN PINNIPED PREHISTORY

As noted at the beginning of this chapter, whereas plesiosaurs do at least have a fossil record substantiating their erstwhile existence, there is no evidence whatsoever in the currently-known fossil record for the existence at any time in pinniped history of an extreme, veritable giraffe-necked form like *Megalotaria* as predicted by Heuvelmans *et al.* as the identity of aquatic long-necks. Indeed, the only confirmed evidence for the former existence of *any* seals possessing necks that were in any way longer than those of modern-day species is the series of fossil remains from the so-called swan-necked seals belonging to the extinct phocid genus *Acrophoca*.

But just how long *were* their necks, and were they long enough to justify their popular 'swan-necked' tag? Dating from the late Miocene to early Pliocene (approximately 7-4 million years ago), the first species to be discovered and named was *Acrophoca longirostris*, which was formally described by palaeontologist Dr Christian de Muizon in 1981, and whose fossils have been uncovered in Chile and Peru. It measured up to 5 ft long, and in Muizon's description he revealed that both the length of its cervical vertebrae and the total length of its cervical column exceeded those of all modern-day seals. Moreover, its cervical column length was approximately 21 per cent of its total vertebral column length, whereas in modern-day seals it is generally 17-19 per cent. Its skull was also lengthy (hence its species name, *longirostris*).

Yet although the neck of *A. longirostris* was proportionately longer, it was not as streamlined as the neck of what may well be its closest modern-day relative – Antarctica's leopard seal *Hydrurga leptonyx*, famed for its neck's elongated appearance when striking out at prey (see also below). Moreover, its flippers were less well-developed, a second characteristic indicating that it was less adapted for swimming than the leopard seal, and that it may therefore have spent much of its time around the Pacific's coasts rather than out at sea (a behavioural preference that, if true, has been perpetuated by the leopard seal, in spite of its more specialised form for swimming).

Leopard seal photographed on land in 1910 during the Terra Nova (British Antarctic) Expedition 1910-1912 (public domain)

As for its 'swan-necked' appellation: in a *Tetrapod Zoology* blog article of 4 February 2006 dealing with *Acrophoca*, Dr Darren Naish stated that because the necks of seals are sufficiently flexible to exhibit a marked lengthening effect when they lunge at prey, stretch, or spy-hop:

> ...when alive, *Acrophoca* would have been capable of looking even longer in the neck than we might think just from its fossils. But clearly it's a stretch [pun intended?!] to imagine this animal as having a long long long neck like a swan, or a plesiosaur, so, sadly, 'swan-necked seal' really is a bit of an exaggeration.

In 2002, with fellow palaeontologist Dr Stig A. Walsh, Naish co-described what appeared to be a new, second *Acrophoca* species, based upon fossils retrieved in Chile, but they declined to give it a formal scientific name. This was because substantial new fossil material hailing from Peru suggested the presence of several additional *Acrophoca* species, so it was felt best to await their full description first. Interestingly, one of these new species had an even longer skull than *A. longirostris*, so it may have looked more unusual than the latter.

HIDDEN-NECK LONG-NECKED SEALS – A LITTLE-KNOWN PARADOX

Ironically, however, we do not even have to look back into prehistory to uncover bona fide long-necked seals. So far, this chapter has been assessing attempts by various cryptozoologists down through the ages to propose as the identity of long-neck aquatic cryptids the existence of a highly-specialised species of seal whose defining characteristic is its long neck. In reality, however, what

An adult bull specimen of the northern fur seal (public domain)

is not readily realised is that science has already confirmed the existence of several such species – species, moreover, which are actually alive today. But how can this be? Allow me to explain.

With the notable exception of the leopard seal's well-delineated neck, in most modern-day seal species the neck is largely hidden, often concealed by blubber, to the point of seeming to be all but non-existent in certain forms. A close examination of such species' skeletons, conversely, reveals a very different – and extremely surprising – picture.

On 15 March 2013, American biologist Cameron A. McCormick's blog *Biological Marginalia* posted a fascinating article entitled 'The hidden necks of seals', containing a table of measurements obtained from a range of different pinniped species. For each species, the length of its neck was given as a percentage of the combined length of its thoracic and lumbar (T-L) vertebrae, and the results were quite remarkable to read. Using this comparison, the bearded seal *Erignathus barbatus* had the shortest neck among phocids, at only 21 per cent T-L, whereas the harp seal *Pagophilus groenlandicus* boasted the longest neck, at 35 per cent T-L – exceeding even the leopard seal's 29 per cent T-L. As for otariids, the shortest neck was that of the Australian sea-lion *Neophoca cinerea* at 34.5 per cent T-L, and the longest was that of the northern fur seal *Callorhinus ursinus* at 41 per cent T-L.

But what was most significant was that even the shortest necks were actually much longer than they outwardly appeared to be in the living animal. So in a very real sense, some already-known, modern-day seal species are actually cryptic long-necked seals, or, to be precise, hidden-neck long-necked seals.

In view of this unexpected revelation, one can scarcely even begin to guess at what the neck percentage T-L value might be for a giraffe-necked, *Megalotaria*-type of long-necked seal – especially when we take into account (judging at least from the above data) that there may be an additional neck portion hidden from sight beneath blubber at its basal region. In fact, such an exceptionally long neck could well be of truly plesiosaurian proportions!

THE SEAL(S) OF APPROVAL

Prior to the establishment of the *Journal of Cryptozoology* in 2012, the appearance in a peer-reviewed academic journal of a paper dealing with cryptids was probably just as rare as the beasts documented in it. This is why, back in late 2008 (and in June 2009 online), the publication by the mainstream scientific journal *Historical Biology* of a paper contemplating the possible existence of still-undiscovered pinniped species was of particular note – and, one hopes, an indication of increasing mainstream approval for serious cryptozoological research.

Authored by Drs Darren Naish and Michael A. Woodley (the latter being a Royal Holloway, University of London postgraduate biology student at that time), both with well known cryptozoological interests, together with Royal Holloway computer scientist Dr Hugh P. Shanahan, it was entitled 'How many extant pinniped species remain to be described?'. In it, the authors examined the description record of the pinnipeds using non-linear and logistic regression models in an attempt to ascertain the number of still-undescribed species, and they combined that work with an evaluation of cryptozoological data, featuring such marine cryptids of alleged pinniped identity as the long-neck sea serpent, the merhorse, Vancouver's serpentiform Cadborosaurus, and the Bering Sea's leopard seal-like tizheruk.

From the results obtained, they revealed that three possibly new, currently undescribed species of pinniped match their statistical expectations, but even these, the authors felt, would need to possess some exceptional characteristics if they do indeed exist.

A giraffe-proportioned neck combined with huge body size would certainly be exceptional, but for all the reasons presented and assessed in this chapter, it seems to me at least that these would be highly improbable characteristics for a seal species to possess and yet remain undiscovered by science, especially if it did indeed occur in both marine and freshwater habitats. Consequently, I am not expecting to witness the formal scientific discovery of a *Megalotaria*-type pinniped any day soon – but how I would love to be proved wrong!

AND FINALLY – THE ONE THAT WON'T GO AWAY

As readers of this chapter will no doubt have realised by now, I am definitely not a proponent of the giraffe-necked, *Megalotaria*-type giant pinniped as an identity for any aquatic cryptid. Consequently, I would like nothing more than to jettison it as far away from my thoughts as possible when reviewing such creatures, but there is one tantalising case that always prevents me from doing so – and this is it.

The Orkney Islands and Caithness on the mainland of northern Scotland are separated by a strait of seawater known as the Pentland Firth, which is a popular habitat for seals throughout the year. Two species are known to occur here, the common seal and the grey seal *Halichoerus grypus* – but at about 9.30 am on or around 5 August 1919, off the Orkney island of Hoy, what seems to have been a third, and dramatically different, seal species also made an appearance in this strait, to the astonishment of its eyewitnesses. These consisted of a holidaying lawyer named J. Mackintosh Bell and some local cod fishermen friends of his whom he had chosen to work with on their boat while visiting the Orkneys. His friends had seen the creature before, were very perplexed as to what it might be, and had actually just begun to tell him about it in the hope that he may be able to identify what it was when the subject of their conversation abruptly appeared, not far away from the boat that they were in.

Lieutenant-Commander Rupert T. Gould of Britain's Royal Navy investigated and documented aquatic monsters in his spare time, and after learning about this sighting he contacted Bell and asked him for full details. Bell duly forwarded an in-depth account, which Gould later published in slightly abbreviated form within his book *The Case For the Sea-Serpent* (1930). Four years later, moreover, Gould wrote the first comprehensive study of Nessie, entitled *The Loch Ness Monster and Others*, spending several days at the loch, travelling around it on his motorbike and interviewing many eyewitnesses during his researches for this book.

As far as I am aware, Bell's original, full-length account has never appeared in print, but here is the slightly abbreviated version of it that Gould published in his sea serpent book:

Artistic representation of Cadborosaurus as an exceedingly serpentiform pinniped-like cryptid (© Richard Svensson)

The very first day I was there, I think it was about 5 August, I went afloat with a crew of four at about 9.30 a.m. for the purpose of firstly lifting lobster creels and then for cod fishing. On making our way to the creels, which had been set in a line between Brims Ness and Tor Ness, my friends said "We wonder if we will see that sea monster which we often see, and perhaps you will be able to tell us what it is."

We got to the creels, hauled some, and were moving slowly with the motor to another, when my friends said very quietly "There he is."

I looked, and sure enough about 25—30 yards from the boat a long neck as thick as an elephant's fore leg, all rough-looking like an elephant's hide, was sticking up. On top of this was the head which was much smaller in proportion, but of same colour. The head was like that of a dog, coming sharp to the nose. The eye was black and small, and the whiskers were black. The neck, I should say, stuck about 5-6 ft., possibly more, out of the water.

The animal was very shy, and kept pushing its head up then pulling it down, but never going quite out of sight. The body I could not then see. Then it dis-appeared, and I said "If it comes again I'll take a snapshot of it." Sure enough it did come and I took as I thought a snap of it, but on looking at the camera shutter, I found it had not closed owing to its being swollen, so I did not get a photo. I then said "I'll shoot it" (with my .303 rifle) but the skipper would not hear of it in case I wounded it, and it might attack us.

It disappeared, and as was its custom swam close alongside the boat about 10 feet down. We all saw it plainly, my friends remarking that they had seen it many times swimming just the same way after it had shown itself on the surface. My friends told me that they had seen it the year before just about the same place. It was a common occurrence, so they said. 'That year (1919) was the last of several years in which they saw it annually. It did not show itself again for two or three years, and then it was only seen once. As to its body, it was, seen below the water, dark brown, getting slightly lighter as it got to the outer edge, then at the edge appeared to be almost grey. It had two paddles or fins on its sides and two at its stern. My friends thought it would weigh 2 or 3 tons, some thinking 4 to 6. Not only my friends, but others, lobster fishing, got many chances of seeing it...

I may say that since 1919 all cod and other deep-sea coarse fish have left the Pentland Firth. I think the reason is that such monsters frequent the rocky caves, which are always covered by deep water. My friends think the animal may have been killed by a passing steamer, but I think it is possibly a native of warmer seas, and that if we get a really hot summer it will be seen again.

Bell also furnished Gould with two sketches that he had drawn of the animal (one showing its head and very long neck, the other showing how it looked when swimming underwater), plus a map of the approximate location where they had seen it. This was on the northern side of the Pentland Firth, roughly 1.6 miles north-westward of Tor Ness, the southern point of the Orkney island of Hoy, and about an eighth of a mile offshore, in some 20 fathoms of water.

When Gould wrote to Bell requesting the approximate dimensions of the creature, Bell provided the following additional details:

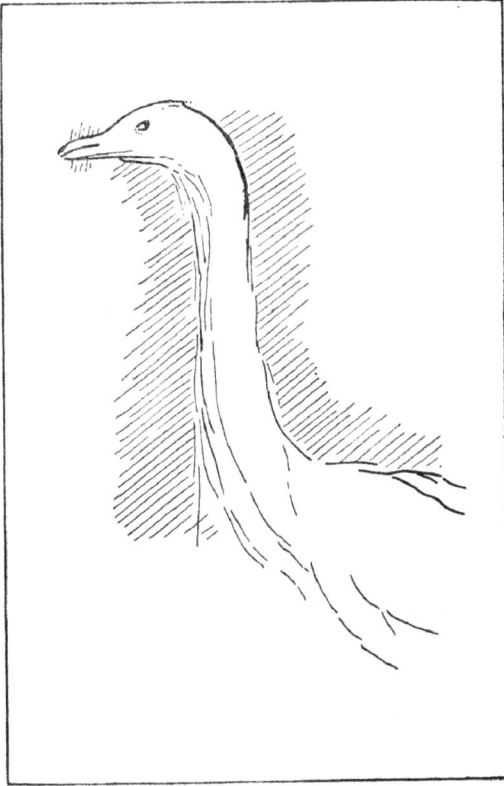

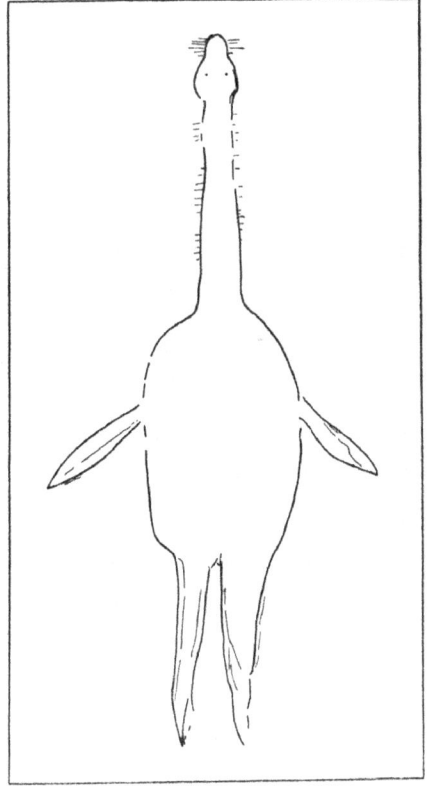

Bell's two sketches of the long-neck that he had seen (public domain)

...Dimensions. Neck, so far as seen, say 6—7 feet. *Body* never seen when neck straight up, but just covered by the water. You could detect the paddles causing the water to ripple. When under water, swimming, the body, I think, to the end of the tail flappers would be about 12 ft. long - and, if the neck were stretched to say 8ft., the neck and body 18—20 ft. long. The skipper of the boat remarked that sometimes the top of the head, when seen from a boat vertically, was a bright red. *Neck* thickness say 1 foot diameter : *Head* very like a black retriever — say 6" long by 4" broad. Whiskers black and short. *Circumference* of body say 10-11 feet, but this I am not sure of, as I never saw all round it, but it would be 4-5 ft. across the back...

Needless to say, everything about this creature, both in Bell's verbal accounts and in his sketches, screams out "Seal!!" – very long neck notwithstanding.

When documenting it in his 2007 review of the long-necked seal concept, Robert Cornes stated: "If this account is true and there appears no reason to think otherwise, then it is arguably the most convincing for the existence of a seal with a long neck". Indeed it is, because if Bell's testimony and sketches *are* accurate, it is difficult to comprehend how the creature that he and his friends saw could have been anything other than a seal – and an exceptionally, extraordinarily long-necked one at that. It is for this reason, if for no other, that the concept of the long-necked seal, even in its most bizarre, giraffe-necked manifestation, continues to frustrate and fascinate me in equal measure, and seems destined to do so for a long time to come.

POSTSCRIPT: FLESHING OUT NESSIE?

Sceptics regularly dismiss Nessie on the grounds that there is no tangible, physical evidence for such a creature's existence, evidence that could be subjected to formal scientific examination in order to determine its originator's taxonomic identity. On one very notable occasion, however, some such evidence was indeed obtained, nothing less, in fact, than sizeable samples of flesh from an apparent Nessie - only for them to be carelessly thrown away! Here's what happened.

In 1978, a holiday cruiser owned by truck driver Stan Roberts, rented out to a family that included an elderly grandfather, collided heavily with a substantial unknown object while sailing on Loch Ness near Urquhart Castle. As later recalled by Roberts in a *Daily Record* interview:

> The propeller stopped turning. The family were very alarmed. The old man had a heart attack and seemed to have died. There was no radio on board so they let off distress flares to get a tow back to Fort Augustus. The grandfather was taken by ambulance to hospital where he was found to be dead.

Roberts was duly informed by the rental managers. However:

> They simply told me there had been an accident. It was only later that I learned more - what had been found on the underside of the boat when they pulled it out of the water.

Boatyard workers who examined the cruiser had found:

> ...flesh and black skin an inch thick along the propshaft". [However,] the workers chiseled the flesh away and threw it into the Caledonian Canal. I said you stupid b-----s. It would have proved that Nessie was here.

Indeed it might. Certainly, to quote Adrian Shine of the Loch Ness Project when told of this incident:

> Very frustrating. With modern DNA techniques we could have learned a lot about exactly what had caused the damage.

In fact, this was quite possibly the single greatest lost opportunity in the entire LNM history to conduct a direct scientific examination and identification of Nessie, because there is no *known* animal species resident in the loch that is big enough to have caused such a collision.

CHAPTER 4:
DATING NESSIE

How an aquatic monster was driven off by virtue of the blessed man's prayer

On another occasion also, when the blessed man was living for some days in the province of the Picts, he was obliged to cross the river Nesa (the Ness); and when he reached the bank of the river, he saw some of the inhabitants burying an unfortunate man, who, according to the account of those who were burying him, was a short time before seized, as he was swimming, and bitten most severely by a monster that lived in the water; his wretched body was, though too late, taken out with a hook, by those who came to his assistance in a boat. The blessed man, on hearing this, was so far from being dismayed, that he directed one of his companions to swim over and row across the coble that was moored at the farther bank. And Lugne Mocumin hearing the command of the excellent man, obeyed without the least delay, taking off all his clothes, except his tunic, and leaping into the water. But the monster, which, so far from being satiated, was only roused for more prey, was lying at the bottom of the stream, and when it felt the water disturbed above by the man swimming, suddenly rushed out, and, giving an awful roar, darted after him, with its mouth wide open, as the man swam in the middle of the stream. Then the blessed man observing this, raised his holy hand, while all the rest, brethren as well as strangers, were stupefied with terror, and, invoking the name of God, formed the saving sign of the cross in the air, and commanded the ferocious monster, saying, 'Thou shalt go no further, nor touch the man; go back with all speed.' Then at the voice of the saint, the monster was terrified, and fled more quickly than if it had been pulled back with ropes, though it had just got so near to Lugne, as he swam, that there was not more than the length of a spear-staff between the man and the beast. Then the brethren seeing that the monster had gone back, and that their comrade Lugne returned to them in the boat safe and sound, were struck with admiration, and gave glory to God in the blessed man. And even the barbarous heathens, who were present, were forced by the greatness of this miracle, which they themselves had seen, to magnify the God of the Christians.

Adomnán – *Life of St Columba* (English translation)

In the opening paragraph of my account dealing with the Loch Ness monster within my book *In Search of Prehistoric Survivors* (1995), I stated:

Its distinguished history supposedly dates back at least to the sixth century AD, when a mysterious water beast was encountered by St Columba (though in River Ness, not Loch Ness, and at about 580 AD, rather than 565 AD as given in many other works).

Since then, I have received a number of communications from readers querying this statement of mine regarding the date of the encounter, and questioning whether I was mistaken, bearing in mind (as I had myself stated) that so many other publications do indeed cite 565 AD as the date in question. (The earliest report of this encounter, incidentally, can be found in the *Life of St Columba*, written in Latin by Adomnán, a subsequently canonised abbot of Iona Abbey, and published during the 7th Century – the relevant section in translated form is the opening quote of this present chapter.)

As I have recently received yet another query from a reader concerning this matter, I would like to set the record straight once and for all.

I am not mistaken. On the contrary: rather than being content merely to repeat uncritically the date given by previous cryptozoological writers in their own works, I purposefully utilised as my source of information regarding this particular issue an authoritative scientific paper whose author, to quote his own words: "...happen[s] to work professionally in the appropriate field of insular Celtic scholarship".

The paper in question, published in 1988 by the now-defunct International Society of Cryptozoology's scholarly, refereed journal *Cryptozoology* (vol. 7, pp. 38-45), was entitled 'The "Monster" Episode in Adomnan's Life of St. Columba', and was authored by Prof. Charles Thomas, Director of Exeter University's Institute of Cornish Studies. As an expert in Celtic history, Prof. Thomas was eminently qualified to comment upon when - and where - St Columba famously encountered the mysterious water beast. So what did he say about it?

Within his paper, Thomas announced that this happened probably around 580 AD (rather than 565 AD). Of even greater significance, Thomas also exposed another popular fallacy - revealing that the encounter did not actually occur at Loch Ness. Instead, it took place on the River Ness, and he suggested that this particular water beast may not have had any connection, therefore, with the mystery creature(s) reported in later times from Loch Ness.

In fact, it might simply have been a straggler belonging to a scientifically-known but locally-unfamiliar species - a rare visitor from Arctic zones, perhaps, such as a walrus or a bearded seal. This is one reason why, when documenting Nessie in my book, I included the word 'supposedly' in the phrase "Its distinguished history supposedly dates back...".

Of course, it is possible that one of the Nessie-type beasts from Loch Ness found its way into the river. Nevertheless, without any additional data to examine, it is evident that this most famous alleged sighting of Nessie in early times is not as clear-cut as many writers, and readers, have traditionally - but erroneously - assumed it to be.

THE 'OTHER' ENCOUNTER BETWEEN A BRITISH SAINT AND A BRITISH LAKE MONSTER

Incidentally, whereas every self-respecting cryptozoologist knows – or should know – about the famous encounter claimed for St Columba and the Loch Ness monster during the 6th Century AD as referred to above, far less famous, conversely, yet no less interesting, is the 'other' encounter between a British saint and a British lake monster (by which I mean a saint

St Columba confronts the monster (© William M. Rebsamen)

St Mochua and the lake monster (© William M. Rebsamen)

and a lake monster from the British Isles, a geographical term embracing the major islands of Great Britain and Ireland plus the many smaller islands collectively surrounding them). Without further ado, therefore, here is this little-known latter incident's intriguing history.

Born Crónán mac Bécáin sometime during the late 6th Century or early 7th Century AD (his age upon death is unrecorded but he is known to have died on 30 March 637 AD), St Mochua as he became was a notable early Irish saint. After completing his education at Bangor, he was reputedly led through many regions of Ireland by a miraculous fountain that could move of its own accord, and during that time he performed a considerable number of miracles himself. One of these involved a lake monster.

At one point during his fountain-guided travels, Mochua crossed the River Shannon and entered the territory of Omania (aka Hy-Many) in the province of Connaught, where he visited Kellach, who was the son of Ragallus, king of Connaught. During Mochua's stay, Ragallus went out hunting a stag, which in order to escape pursuit, leapt from a steep precipice on the shore of a lake and swam to a large rock, jutting forth above the lake's waters like a small island, where the creature sought sanctuary. Different reports cite different lakes (or loughs) as being the one in question here – including County Galway's Lough Ree, Lough Cime or Cimbe (aka Hackett), and Lake Raminium.

According to traditional legend, Mochua had stated that God would safeguard anyone who was bold enough to dive into the lake and swim after the stag, so with Ragallus's permission one brave man did precisely that, reaching the rock-island and killing the stag. When he attempted to swim back to the lake's shore with his slain quarry's carcase, however, a water monster that lived in this body of water and was held in great dread by the populace on account of its murderous nature suddenly rose up from the depths and promptly devoured the hapless swimmer.

When this happened, Ragallus angrily chastised Mochua, because his prediction had seemingly failed to come true, but in response Mochua confronted the monster and prayed – whereupon the monster abruptly vomited forth the swimmer, who was alive and totally unharmed. Moreover, never again did this great water beast attack any other swimmer. Ragallus and his subjects duly gave thanks to God for this miracle, and Mochua was afterwards greatly revered here.

Whether any such monster truly existed or whether this is just a hagiographical legend is unclear (who knows, it may even be derived directly from the earlier one featuring St Columba). And even if it did exist, there does not appear to be any description of the creature on record that might allow a serious attempt to be made at identifying its species.

Nevertheless, bearing in mind that the loughs of Connaught, including Ree, have featured prominently in modern-day reports of Irish freshwater monsters, notably those of the sinuous horse-eel variety, it is an intriguing story, and perhaps even one of bona fide cryptozoological relevance and precedence.

A zeuglodont whale (bottom left) and a plesiosaur (top right), two popular Nessie identities discussed during the Loch Ness Symposium at the International Society of Cryptozoology's Annual Membership Meeting held in July 1987 at Edinburgh's Royal Museum of Scotland (© Richard Svensson)

CHAPTER 5:

IN CONFERENCE WITH NESSIE AT EDINBURGH

The 1987 Membership Meeting was certainly the most successful the Society has ever held. Furthermore, it is believed to be the first time that a scientific institution [the Royal Museum of Scotland] has hosted a meeting on the Loch Ness Monster (a planned conference at the University of Edinburgh in the 1970's [sic] was cancelled because of premature and sensational publicity).

J. Richard Greenwell – *ISC Newsletter*, Winter 1987

In 1982, cryptozoology took a momentous step forward, with the establishment of the International Society of Cryptozoology (ISC), which was the world's first scientific society devoted to the investigation of animals whose existence is currently not officially recognised by science (sadly, the ISC is now defunct). Cryptozoology's further advancement towards full acceptance was greatly assisted by the ISC's policy of staging an Annual Membership Meeting, held each year at a different scientific institution and attracting considerable professional and public interest. In 1985, the ISC also sponsored a one-day cryptozoological symposium contained within the Third International Congress of Systematic and Evolutionary Biology, held at the University of Sussex, in Brighton, which marked the ISC's first visit to the United Kingdom.

Two years later, however, July 1987 saw the first ISC Annual Membership Meeting to be staged in the UK. Moreover, this was also the first two-day Membership Meeting held by the Society, and the first in which the presentations were grouped thematically. In addition, by special accord the gathering on this particular occasion took the form of a joint meeting - of the ISC and the Scottish Branch of the SHNH (the Society for the History of Natural History), whose base is London's Natural History Museum.

The meeting was held at Edinburgh's auspicious Royal Museum of Scotland. In addition to ISC and SHNH members, for a nominal fee of £1.00 non-members were also admitted. The symposia were chaired by the museum's then Curator of Mollusca, Mr David Heppell, who also served at that time on the ISC's Board of Directors.

Day Two's symposium was devoted to cryptozoological cats, in which, as one of several participating speakers, I presented a paper on the origin and possible zoological identity of the Kellas cat. Other papers dealt with British mystery cats, the king cheetah, the onza, and the Queensland tiger. Day One's symposium, conversely, which I also attended and is the subject of this chapter, was devoted entirely to the world's most famous mystery beast – the Loch Ness monster.

SEARCH FOR NESSIE
At 10.00 am on 25 July, Day 1's symposium formally commenced. It was entitled 'The Search For Nessie in the 1980s', and was officially launched by Dr Robert G. Anderson, Director of the National Museums of Scotland, who welcomed the societies and the audience to the museum. He dedicated the meeting to the memory of two persons who were noted for their keen cryptozoological interests - the late David James (Honorary Member of the ISC and co-founder of the Loch Ness Investigation Bureau - LNIB), and the late Ian Lyster (Curator of Ornithology at the Royal Museum of Scotland).

A very impressive plesiosaurian Nessie model at the Loch Ness Exhibition Centre, Drumnadrochit, photographed by me during my own visit there following the Nessie symposium (© Dr Karl Shuker)

The first paper, entitled 'The History of the Loch Ness Monster', was presented by Dr Richard Fitter – then Chairman of the Fauna and Flora Preservation Society (FFPS) and also a co-founder of the LNIB (which functioned from 1962 to 1976). Dr Fitter recounted a concise history of the Loch Ness phenomenon, from the days of St. Columba's sightings of the monster during the 6th Century AD, through the resurgence of Loch Ness interest in the early 1930s, onwards to the LNIB's work during the 1960s, and into the 1980s and recent studies, complementing his presentation with a 12-minute film.

After the fundamental Nessie question: "Is there a Loch Ness monster?", the next most-repeated query must surely be: "What is the Loch Ness monster?" This latter subject was dealt with comprehensively by Prof. Roy P. Mackal, a prominent biochemist, cryptozoologist (he was the ISC's Vice-President), and longstanding Loch Ness investigator. In his paper, 'The Biology of the Loch Ness Monster', Prof. Mackal analysed the morphology and physiology of each group of animals put forward in the past as identities for Nessie. He concluded that mammalian or reptilian identities were the most likely candidates, with amphibian or soft-bodied invertebrate suggestions amongst the least plausible.

Mackal also spoke about a possibility of obtaining evidence that would conclusively identify at least one North American version of Nessie. For he noted that in Canada, fishermen have reported to him that they often see such creatures following the salmon swimming upstream in rivers to spawn. Mackal suggested that if nets were stretched across one such river at the time when the salmon appear, it may actually be possible to snare one of these Nessie-type beasts! Needless to say, such an acquisition would constitute a tremendous zoological discovery, and it is to be hoped that such a promising venture will indeed take place.

The next paper, 'Public Perceptions of the Loch Ness Monster', was presented by Dr Henry H. Bauer, Professor of Chemistry at Virginia Polytechnic Institute and State University. Dr Bauer focussed his attention not upon the monster itself or upon monster research findings, but instead upon the sociological, philosophical, and psychological components of these subjects. The result was a most thought-provoking presentation, discussing the nature and features of belief and disbelief using Nessie as the example, and examining the ways in which these have varied during the long history of Loch Ness reports and investigations.

IMPRESSIVE EVIDENCE

The most famous and contentious photo purportedly of Nessie ever obtained is certainly the popularly-termed 'Surgeon's Photograph', snapped in April 1934 by London gynaecologist Robert Kenneth Wilson, and depicting a black object resembling a slender neck surmounted by a small head extending above the rippled water surface. In his paper entitled 'The Wilson Nessie Photo: A Size Determination Based on Physical Principles', Prof. Paul LeBlond, an oceanographer at the University of British Columbia, demonstrated that a size estimate of the object in Wilson's photograph can be obtained by relating the appearance (in terms of surface disturbance and wave formation depicted in the photo) of Loch Ness's surface to wind speed and thence to wind waves' lengths. As a result of such comparisons, an estimate of 4 ft was obtained for the height of the object above the water surface. This is a principle which Prof.

Portrait of Adrian Shine in the company of Nessie and a water-horse of the classical Greek hippocampus variety (© Thomas Finley)

LeBlond had already applied to the equally controversial Mansi photograph of a creature-like object on Lake Champlain, and will be of great benefit to future aquatic monster research.

Following lunch, the presentations continued with a paper entitled 'Recent Fieldwork by the Loch Ness and Morar Project', presented by Adrian Shine, the Project's Field Leader. In 1974, the British-instigated Loch Morar Expeditions began, and were succeeded by the Project, which concentrated thereafter upon Loch Ness. Much of its efforts were directed towards sonar/echo-sounding investigations, and Shine described the more recent work in this line carried out by the Project, plus a new survey involving an extensive multi-craft, underwater sonar sweep along the loch, in an attempt to detect the presence of any large creatures that may exist there.

Certainly, the most visually impressive (but again highly controversial) evidence for the existence of such beasts to have been procured via underwater photography consists of the 'flipper' photographs. These were obtained in 1972 by the Loch Ness research team from the USA's Academy of Applied Science, headed by its president Dr Robert H. Rines. Consequently, in his paper 'A Review of Research Contributions to Date of the Academy of

Applied Science at Loch Ness', Dr Rines discussed in detail these photographs (and the criticisms that have been levelled at them by various armchair Nessie sceptics), together with the further, equally intriguing, underwater photographs obtained by the Academy's team in 1975. One of these latter pictures was thought by some to feature the head and neck of a large

When this photograph of Loch Ness – snapped by me during my visit there following the symposium – was developed, a blob could be plainly seen apparently at the loch's surface in the far distance. It may simply be an artefact of the photographic development process, or merely a boat whose presence had not attracted my attention when taking the photo – or it might just be Nessie! (© Dr Karl Shuker)

creature, another resembled a close-up of a creature's head (the so-called 'gargoyle' photograph). Moreover, he ended his paper with the tantalising statement that his team now had access to recently-declassified equipment that should render the loch transparent (figuratively speaking, that is) as far as future searches and observations appertaining to Loch Ness's mysterious denizens were concerned.

CLASSIC FILM

In a special announcement following Dr Rines's presentation, the ISC formally honoured the final speaker at this symposium - aero-engineer, author, and world-renowned Loch Ness investigator Tim Dinsdale - for his most distinguished and significant contributions to Loch

Ness monster research. This was well-deserved recognition for the sterling work of this most diligent, courteous, and respected cryptozoological researcher. (Tragically, however, less than five months later he suffered a fatal heart attack.)

In his paper, entitled 'Three Decades of Nessie Hunting: A Personal Odyssey', Tim Dinsdale traversed through his many memorable years of very impressive and fascinating personal investigations of the Loch Ness phenomenon, which began in earnest with what must surely have been the most exhilarating moment in his entire three-decade quest. This occurred on 23 April 1960, when he obtained his classic film of a huge creature-like object, partially visible above the water surface, swimming rapidly across the loch. This electrifying event served as a source of great motivation for him during his subsequent researches, detailed within his presentation, which continued each year since then. Also referred to alongside these were his involvements with many of the investigations that had at some stage or another been carried out by other researchers at the loch from 1960 onwards, plus the wide diversity of equipment, techniques, and vehicles that he has utilised in his examination and exploration of one of cryptozoology's most endearing and enduring of enigmas.

Following Tim Dinsdale's presentation, a panel debate took place, during which an interesting exchange of questions, replies, and opinions occurred between the audience and the panel – the latter consisting of all of the Loch Ness symposium's speakers.

At the end of Day Two, following David Heppell's final remarks as Chairman the joint meeting of the ISC and SHNH was formally adjourned - but all was not completely over. A small party of us set off for a day trip to Loch Ness itself. Sadly Nessie did not make an appearance...or did she?

However, we were able to visit the notable Loch Ness Exhibition Centre at Drumnadrochit. This contained a veritable cornucopia of Nessie information and exhibits - ranging from analyses of all of the major surface and underwater monster photographs obtained to date, selections of sonar evidence recorded, and models of the loch itself plus the various animal forms that have been offered as possible Nessie identities, to living examples of some of the loch's smaller inhabitants, exhibits of various vehicles and pieces of scientific equipment used in Loch Ness researches, and much more. It even boasted a monstrously-large plesiosaurian Nessie replica gazing intently at the Centre's visitors from its very own adjacent pool - just in case the real star failed to give a performance.

NOTABLE SUCCESS
From the size of the audience and the considerable interest that it engendered among the general public and the media, as well as among the scientific fraternity, this Sixth ISC Annual Membership Meeting had evidently been more than just a notable success in itself. It had also constituted a most favourable and significant contribution to cryptozoology's continuing emergence as a respectable and respected scientific discipline. Moreover, its presentations had underlined very effectively the tremendous impact upon science that cryptozoological discoveries do (and will continue to) make.

CHAPTER 6:

A HITHERT⊕-UNPUBLISHED SIGHTING ⊕F THE L⊕CH NESS M⊕NSTER?

Tim Dinsdale, the adventurer and Loch Ness hunter, lives on in the Tim Dinsdale Memorial Award.

The Dinsdale Award was established in 1992 by The Society for Scientific Exploration's founding member, councilor, and editor of the SSE Journal, (now retired) Professor Henry Bauer, so that the SSE could recognize "significant contributions to the expansion of human understanding through the study of unexplained phenomena."

Such contributions were made by Tim Dinsdale, in whose memory this prize is named. Dinsdale was by profession an engineer, who chanced to obtain in 1960 what remains still the most striking evidence of unexplained animals in Loch Ness. Dinsdale's subsequent investigations over three decades were carried on with such integrity that the *Times* of London marked his passing with a respectful obituary, rare indeed with someone whose prominence stems from the pursuit of such unorthodox research.

Through the Dinsdale award, the SSE endeavors to identify, publicize, and reward senior scholars who have made similarly substantial contributions to the understanding of anomalous physical, biological, and psychological events in the spirit of meticulous research, exemplary methodology, and proper scholarly attitude that Tim Dinsdale exemplified.

Loren Coleman – *Cryptomundo*, 1 July 2008

Back in 1986, veteran Loch Ness monster researcher Tim Dinsdale (1924-1987) had corresponded with me regarding a water monster very different from Nessie but no less intriguing, and which formed the basis of my first major cryptozoological investigation. This was Gambo, the Gambian sea serpent.

Consequently, when I attended the International Society of Cryptozoology's two-day Membership Meeting held at the Royal Museum of Scotland in Edinburgh on 25-26 July 1987, Day 1 of which was devoted to Nessie and at which Tim was one of the speakers (see this present book's previous chapter to read my reporting of the meeting), I lost no time in introducing myself to him so that I could thank him directly for his kind interest and encouragement in my own fledgling cryptozoological researches.

In response, Tim gave me a copy of a most interesting Nessie sighting that an eyewitness had recently sent to him. He didn't provide me with the eyewitness's name, because he no doubt intended to document it in some future publication. Tragically, however, on 14 December 1987 Tim suffered a fatal heart attack.

Recently, I came upon the copy of this sighting that Tim had given to me all those years ago, and as I was unsure whether it was ever made public, I decided that I should do so, by including in a post of 30 July 2013 for my *ShukerNature* blog the sketches of what the eyewitness claimed to have seen, together with the sparse details concerning it that I have on file, in case this may be of benefit to other Nessie researchers. I am now also publishing them for the first time in hard-copy format:

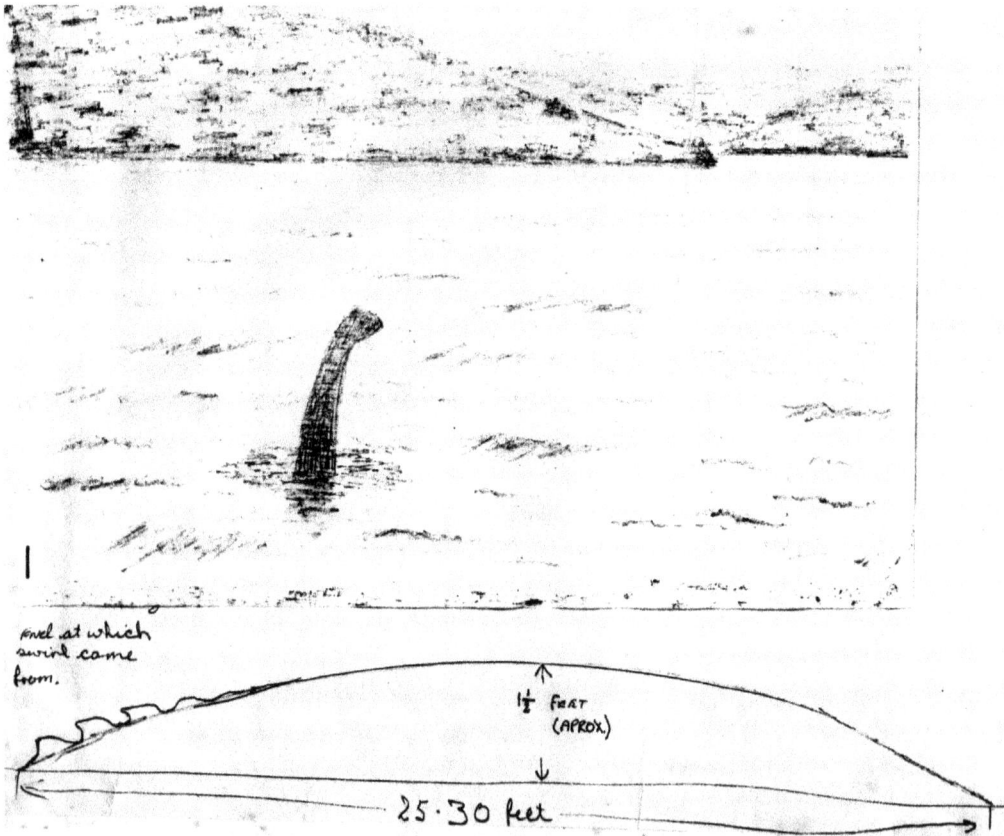

The sheet given to me on 25 July 1987 by Tim Dinsdale containing two sketches by a Nessie eyewitness (© unknown)

The eyewitness observed a typical 'periscope' shape projecting up through the water surface, yielding an outline reminiscent of the object in the controversial Surgeon's Photograph. He/she also saw a very long hump visible above the water surface, approximately 25-30 ft in length and approximately 1.5 ft high, with what looked like distinct backward-pointing serrations running along the posterior portion of its upper surface.

If anyone has any further knowledge concerning this Nessie sighting, I'd be delighted to receive details.

BOTHERSOME BEITHIRS

Speaking of little-publicised sightings: Nessie may well be Scotland's best known freshwater mystery beast, but it is not this country's only one. Far less familiar yet no less intriguing in its own way is the beithir.

In 1994, a correspondent to English magazine *Athene* published two fascinating articles concerning modern-day beithir sightings. During early 1975, he met a fisherman near Inverness who claimed that he and four others once sighted a beithir lying coiled in shallow water close to the edge of a deep gorge upstream of the Falls of Kilmorack. When it realised that it had been observed, however, it thrashed wildly about before finally swimming up the gorge near Beaufort Castle and disappearing. The fishermen estimated its length at around 10 ft.

Four months later, the *Athene* correspondent learnt of another sighting, this time offshore of Eilean Aigas, an island in the River Beauly, Highland. He was also informed by a keeper at Strathmore that during the 1930s his wife's parents had seen beithirs moving overland at Loch a' Mhuillidh, near Glen Strathfarrar and the mountain of Sgurr na Lapaich. After discussing these reports with various zoological colleagues, he considered that the beithir was probably an extra-large variety of eel – fishes that are well known for their ability to leave the water and move overland to forage when circumstances necessitate, and even to sustain themselves out of water for protracted periods.

Indeed, a Devon farmer informed the *Athene* correspondent that during the extremely harsh winter of 1947 his mother had been badly frightened to find several eels alive and well in the farm's hayloft, where they had evidently been sheltering since the freezing over of the nearby river some time earlier. The rest of the family came to see this wonder, including the farmer himself (then still a boy), and his father confirmed that they were indeed eels, and not snakes (as his mother had initially assumed).

A report of what may also have been a beithir, or something very similar, had appeared almost a century earlier in the *Inverness Courier* newspaper. Dated 6 March 1856, the report claimed that the creature had appeared in an inland loch on the northwestern Scottish island of Lewis-with-Harris (the largest member of the Western Isles or Outer Hebrides), and reads as follows:

The Sea Serpent in the Highlands.—The village of Leurbost, parish of Lochs, Lewis, is at present the scene of an unusual occurrence. This is no less than the appearance in one of the inland fresh water lakes of an animal which from its great size and dimensions has not a little puzzled our island naturalists. Some suppose him to be a description of the hitherto mythological water-

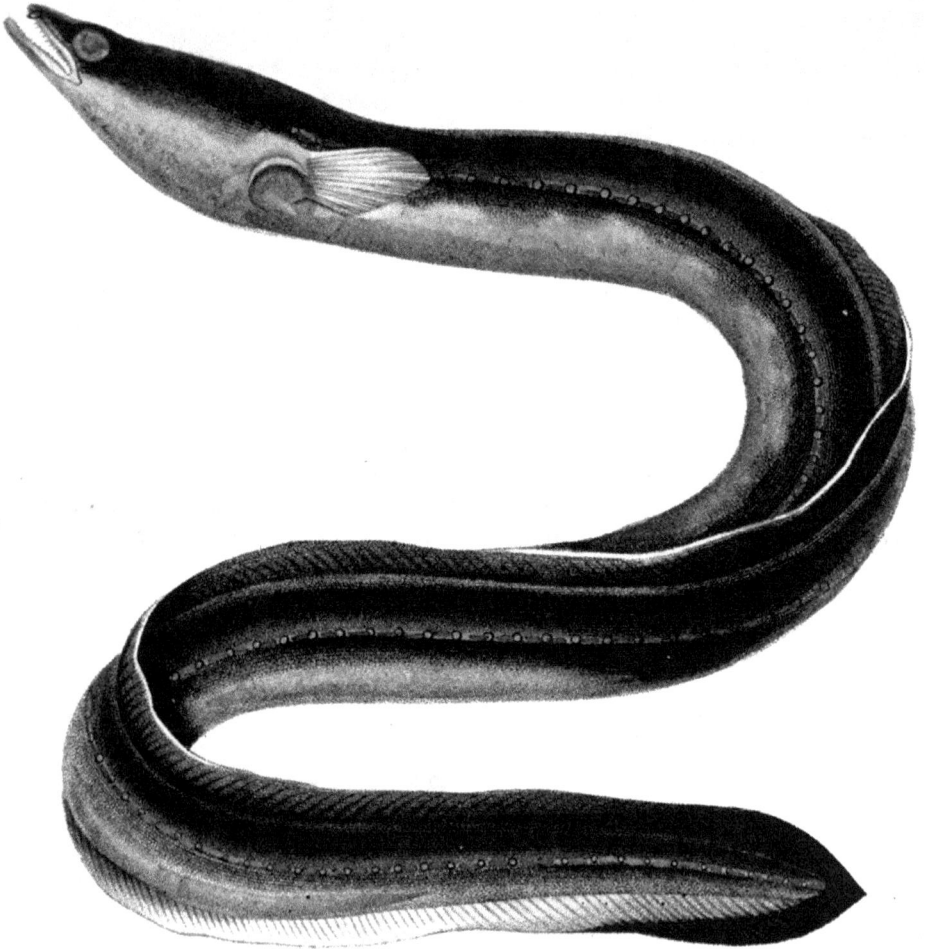

Extra-large European eels *Anguilla anguilla* are plausible candidates for the identity of the beithir (public domain)

kelpie; while others refer it to the minute descriptions of the "sea serpent," which are revived from time to time in newspaper columns. It has been re-peatedly seen within the last fortnight by crowds of people, many of whom have come from the remotest parts of the parish to witness the uncommon spectacle. The animal is described by some as being in appearance and size like a "huge peat stack," while others affirm that a "six-oared boat" could pass between the huge fins, which are occasionally visible. All, however, agree, in describing its form as that of the eel; and we have heard one, whose evidence we can rely upon, state that in length he supposed it to be about 40 feet. It is probable that it is no more than a conger eel after all, animals of this description having been caught in Highland lakes which have attained a huge size. He is currently reported to have swallowed a blanket inadvertently left on the bank of the lake by a girl herding cattle. A sportsman ensconced himself with a rifle in the vicinity of the loch during a whole day, hoping to get a shot, but did no execution.

Bearing in mind that the maximum confirmed length for the European conger eel *Conger conger* is 'only' 10 ft, I cannot help but wonder how seriously one should take any report claiming that specimens of this species which have attained a huge size have indeed been caught in Highland lochs. All I can say is that if they have been, I would very greatly welcome sight of any documented verification of their erstwhile reality.

Nessie as a goat-sized, frog-like monster allegedly spied resting upon an underwater rocky shelf by a diver during the 1870s-1880s
(© Richard Svensson)

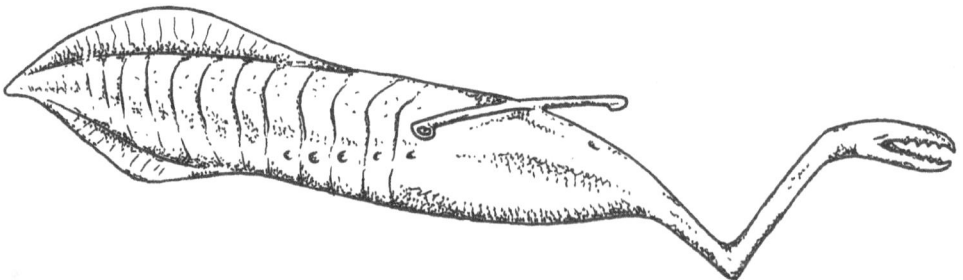

Modern reconstruction of *Tullimonstrum gregarium* in life – a hypothetical giant modern-day version of which was the very unusual Nessie identity championed by F.W. Holiday (© Tim Morris)

CHAPTER 7:

AN UNLIKELINESS OF NESSIES ‐ SURVEYING THE STRANGEST IDENTITIES EVER SUGGESTED FOR THE LOCH NESS MONSTER

Whatever can it be? The simple answer to this question – assuming that one considers it worth the asking – is that no one knows what the Monster is or, more correctly, of what species it may be...

Commander R. T. Gould, in the early 1930s, examined various animal types as possible contenders for the title, and found them wanting: Aquatic birds, salmon, otter, porpoise, tortoise, turtle, catfish, eel, ribbonfish [oarfish], beluga (or any other known species of whale), sturgeon, crocodile, known seals (walrus, sea-lion, sea-leopard, and sea-elephant), shark, sunfish, ray, and giant squid. There is no reason to doubt the conclusions of this brilliant man. Nor is there any reason to accept the alternatives put forward frequently by members of the 'lunatic fringe', of whom there are many at the loch.

The fact remains that until we film the Monster clearly, in good definition, or secure a small piece of tissue from him, or even catch a specimen, the fascinating problem posed by his identity remains, and is open to discussion.

Tim Dinsdale – *Loch Ness Monster* (Fourth Edition)

Today, the classic, pre-eminent image indelibly engrained in everyone's mind when speaking of Nessie is that of a plesiosaur lookalike, complete with long slender neck and tail, small head, and four large diamond-shaped flippers. However, this was not always the case. In the past, a great diversity of alternative ideas concerning the likely appearance and identity of Scotland's cryptozoological megastar existed. Nevertheless, with the exception of just a few (such as a sturgeon, a hypothetical long-necked seal, or various misidentified familiar animals like otters and swimming deer) that still linger tenaciously in the romantic but decidedly plesiosaurian shadow of the general public's favourite concept for Nessie, these other options have largely been forgotten or discarded. Yet they included some truly extraordinary notions and fascinating sightings, which richly deserve their belated resurrection here, as we examine just a selection of those most curious of LNM identities - identities that may have been, might still be, and surely could never, ever be...could they?

OTTERLY MYSTERIOUS

The most familiar cryptozoological identity proffering a furry or hairy mammalian Nessie as opposed to a sleek scaled or scaleless reptilian counterpart is a giant long-necked seal - of the kind famously postulated during the 1960s by pioneering cryptozoologist Dr Bernard Heuvelmans for his 'long-neck' category of sea serpent, and later adopted for lake monsters too by Peter Costello. This identity still receives some attention today (see Chapter 3 for a comprehensive documentation). In contrast, a second mammalian cryptid contender has long been consigned to obscurity – a giant long-necked otter.

Its principal proponent was British zoologist Dr Maurice Burton. Although dismissing most Nessie reports as floating algal mats or misidentified known animals, in his book *The Elusive Monster* (1961) he considered it possible that a small number of reports genuinely featured an undiscovered lutrine form. And perhaps his most memorable claim was that if a long-necked giant otter did exist, it should not be looked for in the loch but on land instead: "...in the marshes or on islands (e.g. Cherry Island [a small island on the loch itself, at Fort Augustus]), up the burns and rivers or along the shores of the loch, although it may also be seen occasionally in the water".

Could Nessie be a giant otter? (public domain)

How ironic it would be if generations of Nessie seekers have been looking for the LNM in entirely the wrong habitat! Intriguingly, an unknown long-necked giant otter-like beast has long been reported from western Ireland, where it is termed the dobhar-chú or master otter (see Chapter 16).

A FROG AS BIG AS A GOAT?

What was an exceedingly unusual sighting of Nessie, made as it was underwater, took place one day during the 1880s-1890s, when diver Duncan Macdonald was lowered into Loch Ness at Johnnie's Point, close to the loch's Fort Augustus entrance to the Caledonian Canal. According to modern-day retellings of this incident (but see this chapter's newly-added postscript for the very different, far less dramatic original version), not long afterwards he resurfaced, gesticulating wildly to his colleagues to pull him out, and in such a terror-stricken state that it took several days before he was finally able to explain why he had been so scared. These graphic details seem to be melodramatic embellishments subsequently added to the original version by person(s) unknown.

What the original version and later retellings of it do agree upon, however, is that MacDonald revealed that while he had been seeking at a depth of around 30 ft the sunken ship that he had been sent down to investigate, he saw a very large, totally unfamiliar creature lying upon a shelf of rock that had been supporting the ship. According to his description, the creature resembled a huge frog, as big as a goat, and it was staring directly at him, albeit not in an aggressive or threatening manner.

The source of this information was Macdonald's own grand-nephew, Donald Fraser, a loch-keeper at Fort Augustus; and it first appeared in print on 31 January 1934, in an article published by the *Northern Chronicle* newspaper. This was the same year that saw the publication of the very first book devoted to Nessie, although it was little more than a pamphlet – entitled *The Home of the Loch Ness Monster* and written by W.H. Lane, it suggested that Nessie may be a giant form of salamander.

Moreover, the concept of Nessie being a gigantic amphibian was revisited four decades later. This was when, in 1976, American cryptozoologist Prof. Roy P. Mackal published what remains the most scientific, rigorously objective study of the LNM in book form.

Entitled *The Monsters of Loch Ness*, in it Prof. Mackal meticulously examined every reasonable zoological identity, and concluded that the most plausible Nessie candidate was a species of giant newt- or salamander-like amphibian, which in his view would account for 88 per cent of the LNM characteristics on file (as opposed to 78 per cent for a species of eel, 69 per cent for a plesiosaur, 59 per cent for a mollusc, 56 per cent for a seal, and 47 per cent for a sea-cow).

Yet despite the convincing and thorough nature of his researches, Mackal's mega-newt theory failed to break the plesiosaur's limpet-like grip upon the imagination of Nessie seekers and the media at large. Moreover, he later abandoned it in favour of either a mammalian or a reptilian identity instead, as declared during his Nessie presentation at the International Society of Cryptozoology's two-day Annual Members Meeting for 1987, held at the Royal Museum of Scotland in Edinburgh (see Chapter 5).

MIGHTY MANTAS AND EUNUCH EELS

Two of the fishiest Nessie identities – zoologically – feature a couple of very different contenders of the piscean persuasion. Cryptozoologists Paul and Lena Bottriell are most famous for their king cheetah researches, but in 1988 they turned their attention briefly to Nessie, and in an exclusive *High Wycombe Star* newspaper interview published on 28 October they offered a new identity for this aquatic cryptid.

Based upon personal sightings of a school of rays seen while snorkelling off Queensland, Australia,

Paul postulated that the LNM may be a very large ray, sporting a series of dorsal fins along its lengthy slender tail (as species such as the electric ray possess), thereby producing the characteristic Nessie humps if protruding through the water surface. He also proposed that its elongate tail could create the familiar 'head and neck' Nessie image if lifted up out of the water (rays do lift their tails in warning displays). Although an ingenious, original idea, the notion of a ray's tail explaining Nessie's head and neck clashes with LNM eyewitness reports that have claimed the head and neck to be unquestionably sentient, actively observing while above the water surface.

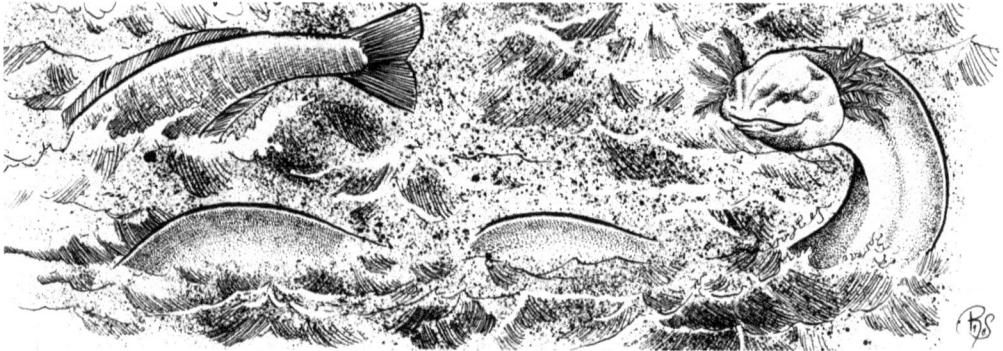

Nessie reconstructed as a giant eel-like fish (© Richard Svensson)

Equally ingenious is a more recent ichthyological identity aired in relation to Nessie. Expanding upon the longstanding belief of various investigators that extra-large eels may be responsible for at least some LNM reports, in 2003 Richard Freeman of the CFZ suggested that Nessie may well be a gigantic eunuch specimen of the common eel *Anguilla anguilla*. That is, one that did not swim out to sea and spawn but instead stayed in the loch, grew exceptionally long (25-30 ft), lived to a much greater age than normal, and was rendered sterile by some currently-undetermined factor present in this and other deep, cold, northern lakes.

I would not be at all surprised to learn that extra-large eels do exist here (indeed, such fishes have been reported by divers in the loch), and they could certainly explain some Nessie sightings of the 'humps above the water' variety. However, I cannot reconcile any kind of eel with the oft-reported vertical above-surface head-and-neck category of LNM sightings, nor with the land sightings that have described a clearly visible four-limbed, long-necked animal (see Chapter 1).

Also, in response to this 'eunuch eel' theory, Dr Scott McNaught, Professor of Lake Biology at Mount Pleasant's Central Michigan University, has stated that even if such eels did arise, they would tend to grow thicker rather than longer. Nevertheless, giant eels remain a distinct possibility in relation to some of the world's more serpentiform lake monsters on record.

IS THE ELEPHANT SQUID A WIZARD IDEA?
Tony 'Doc' Shiels is familiar in the fortean community as a wizard, surrealist artist, showman, and cryptozoological enthusiast, with a particular interest in water monsters, and he claims to have

photographed several, including the LNM. Moreover, in relation to this latter cryptid he has proposed a highly original zoological identity – an as-yet-hypothetical, extremely modified species of enormous squid, which in 1984 he dubbed the elephant squid *Elephanteuthis nnidnidi*.

This was because its most distinctive feature is a long flexible prey-capturing structure resembling an elephant's trunk, which, held above the water surface, would account for Nessie's 'head and long neck' image. Shiels also postulated the presence of inflatable dorsal airsacs for buoyancy purposes, which would explain the many LNM sightings of humps breaking through the loch's water surface.

Naturally, however, so dramatically different a species of squid as this would require a very considerable evolutionary deviation from the more generalised squid blueprint, and not just morphologically. Currently, there is not a single scientifically-confirmed species of freshwater squid on record – every squid presently known to exist today is exclusively marine. So for Shiels's elephant squid to be a reality, it would need to exhibit profound osmoregulatory adaptations to a freshwater lifestyle. For further details concerning this intriguing suggested LNM identity, see Chapter 15.

I'VE NEVER SEEN ANYTHING LIKE IT IN MY LIFE!
The above words, as sung by circus owner Albert Blossom in 20th Century Fox's classic 1967 film musical *Doctor Dolittle* upon first seeing the incredible two-headed pushmi-pullyu, came unbidden but very aptly into my head when, after first reading the thoroughly astonishing, one-of-a-kind Nessie sighting claimed by L. McP. Fordyce, I looked at the accompanying artistic reconstruction based upon his own sketch of what he allegedly saw.

According to his report (*Scots Magazine*, June 1990), his extraordinary encounter occurred in 1932 (just a year before Nessie fever filled the headlines worldwide and the term 'Loch Ness monster' was coined).

Driving along a woodland-surrounded stretch of road leading away from the lochside and towards Fort William, he and his fiancée were amazed to see a huge creature come out of the woods on their left and step over the road about 150 yards ahead towards the loch. Fordyce described it as having:

> ...the gait of an elephant, but looked like a cross between a very large horse and a camel, with a hump on its back and a small head on a long neck...From the rear it looked grey and shaggy. Its long, thin neck gave it the appearance of an elephant with its trunk raised.

He stopped the car, and followed this bizarre animal for a short distance on foot before deciding that it may be safer to abandon his pursuit and go back to his car.

So unlike the typical LNM is this truly weird entity, depicted in Fordyce's account with long slender legs far removed from the flippers more commonly associated with Nessie, that I swiftly checked that this was not the April issue of the magazine in question, but the article ended with an even stranger note. Fordyce revealed that, as stated in Ronald Binns's book, *The*

Reconstruction of Fordyce's truly bizarre 'giraffe-necked camel-horse' Nessie, based upon his description of it (© Tim Morris)

Loch Ness Mystery Solved (1983), in 1771 a Patrick Rose had learnt of a monster seen in Loch Ness that was said to resemble a cross between a horse and a camel. However, this no doubt referred to Nessie's head and neck (rather than to the entire animal), which have indeed been likened to those of a horse on many occasions. So too have those of water monsters elsewhere, including North America's Caddy, a sea monster whose head has been compared with that of a camel as well.

As for Fordyce's giraffe-legged (and –necked) land Nessie (if indeed that is what it was, and not merely some non-native escapee – a female, antler-less moose *Alces alces*, for example, might have been plausible were it not that this species has an exceedingly *large* head, not an extremely small one like the Fordyce creature): always assuming, of course, that this report is genuine, it remains a unique (and uniquely perplexing) curiosity in the LNM annals.

A NOT SO MONSTROUS *TULLIMONSTRUM*?

One of the most bewildering Nessie sightings was that of Mr and Mrs George Spicer. Driving along the road between Dores and Foyers on 22 July 1933, they spied a very large entity emerging from the bushes onto the road ahead. They described it as "an abomination...a loathsome sight", with a long neck, but no apparent limbs, later likened to a massive slug or worm-like beast in some accounts, which lurched rapidly across the road and into the bracken separating it from the lochside.

One Nessie investigator impressed with the prospect of a worm as a suitable explanation was F.W. Holiday. In his book *The Great Orm of Loch Ness* (1969), he nominated a particularly unusual animal as his favoured Nessie. Namely, a hypothetical giant modern-day descendant of a bizarre prehistoric worm called *Tullimonstrum gregarium*, or the Tully monster (after Francis J. Tully, who brought this enigmatic fossil species of uncertain taxonomic affinities to scientific attention in 1955).

What intrigued Holiday about this animal was its unexpectedly Nessie-esque morphology. Unlike more conservative vermiform creatures, *Tullimonstrum* sported a pair of small anterior flipper-like appendages (though these are now known to have been eye-stalks), a pair of cuttlefish-like fins at the rear portion of its body, and a very long, slender jaw-containing proboscis superficially resembling an elongate LNM-type neck and head. However, unlike Nessie, which is often claimed to measure around 30 ft long, *Tullimonstrum* was no more than 1ft, is known only from Illinois, and became extinct over 300 million years ago, during the late Carboniferous Period – all of which no doubt explains why this identity never captured the public imagination.

Interestingly, a team of scientists led by Dr Victoria E. McCoy, formerly of Yale University, publicly announced via a *Nature* paper published online on 16 March 2016 that, following detailed morphological studies of more than 1200 *Tullimonstrum* fossils, using synchrotron elemental mapping and other new analytical techniques, they had discovered that this small but very controversial creature had possessed a notochord. This is a flexible anterioposterior rod composed of cartilage-like material and usually sited dorsally that is a diagnostic feature of all chordates, forming the basis of the backbone. The team also observed the presence of laterally-sited gill

A close-up photograph depicting my resin model of the Tully monster
Tullimonstrum gregarium (© Dr Karl Shuker)

pouches in a few specimens that had been fortuitously preserved lying on their sides during the fossilisation process. Consequently, although it may have been superficially worm-like, *Tullimonstrum* was not any type of invertebrate after all, but a chordate. More specifically, the team consider that it was a vertebrate, and one that belonged to the same lineage as the lampreys.

THE ELEPHANT OF SURPRISE?
As has already been revealed in this chapter, several Nessie identities contain more than an element of surprise, but none more so than when the element is an elephant. After all, whatever Nessie may be, she is certainly no pachyderm...is she?

Remarkably, in a *New Scientist* article of 2 August 1979, Santa Barbara Museum of Natural History director Dr Dennis Power and Illinois University geography research associate Dr

The only portions of an elephant that would be visible above the water surface if it were swimming underwater are those portions present above the horizontal line in the lower illustration here, which, as revealed in the zoomed-in upper illustration, do yield a superficially Nessie-like image – but have elephants ever been swimming in Loch Ness?? (© Dr Karl Shuker)

Donald Johnson cited similarities between the head-and-neck shape in the controversial Surgeon's Photograph of Nessie from 1934 and that of an elephant swimming off the coast of Sri Lanka, and speculated that perhaps travelling circuses have occasionally released their elephants into Loch Ness to bathe, which might then explain the Surgeon's Photograph (an image that in years to come would of course be famously condemned – but never confirmed! – as a hoax). Needless to say, no evidence for any such releases has ever been uncovered.

Nevertheless, as proof that history, even of the strangest variety, can and does repeat itself, in 2006 the startling elephant-Nessie scenario unexpectedly raised its trunk in news headlines all over again - when this very same notion was offered up by palaeontological curator Dr Neil Clark from Glasgow University's Hunterian Museum.

An elephant in the loch's waters, or a giraffe-necked camel-horse in its woods? Bring back *Tullimonstrum* – all is forgiven!

MONSTERS, MONSTERS, EVERYWHERE!
Finally: unlike Nessie's many investigators, R.L. Cassie was one seeker of Scottish Highland water monsters who never had problems finding them. On the contrary, ever since he began his observations, in June 1934, everywhere he looked near his home village of Achanalt he saw monsters! As revealed in this 77-year-old author's mesmerising, self-published two-volume book, *The Monsters of Achanalt* (1935-36), the local rivers and lochs were - at least as far as he could see – quite literally bursting at the seams with monstrous reptiles, and of gargantuan dimensions.

Indeed, one such denizen of Loch Achanalt that he dubbed Gabriel was estimated by him to measure approximately 900 ft long, which meant it was only about 150 ft shorter than the loch itself! Moreover, it was, he claimed, just one of countless other, smaller monsters inhabiting this modestly-sized expanse of freshwater, with plenty more in Lochs Cronn, Culon, Garve, and Rosque - even though most of these are no deeper than 30 ft.

Nor were Cassie's sightings confined to the aquatic domain. As soon as he started looking for monsters on land, where he was convinced that they must breed, he was equally successful - even reporting a sighting of two giant reptilian necks outlined against the snowy face close to the summit of Mòruisg, a prominent 3045-ft-high mountain above Glen Carron in Scotland's northwest Highlands. Not surprisingly, Cassie's absurd observations and books rarely rate a mention in other cryptozoological publications, and are generally dismissed either as the outpourings of an extreme eccentric or as a tongue-in-cheek hoax.

Why, with such a range of other candidates to consider, does the plesiosaur identity steadfastly remain so popular? It is true that, on the one hand, some of the land sightings of Nessie have described an undeniably plesiosaurian entity. On the other hand, however, all manner of scientific objections to the likelihood of a modern-day representative of this officially long-demised lineage of prehistoric aquatic reptile persisting in Loch Ness have been aired over the years. My own belief is that there is no single answer to the mystery of Nessie – instead, I consider it most likely that what we refer to as the LNM is in reality a composite of several different phenomena.

Be that as it may, what seems to raise the plesiosaur's profile far above that of any would-be pretender to the Nessie throne is that the notion of some lingering race of antediluvian monster - an erstwhile contemporary of the mighty dinosaurs, no less - lurking reclusively beneath the loch's dark, mysterious waters conjures forth an incomparably romantic and, equally, chilling scenario that no over-sized newt, emasculated eel, trunk-erecting squid, vermiform wannabe, or even the (very) odd giraffe-necked water camel could ever hope to compete with!

POSTSCRIPT: MORE ABOUT THE 'FROG AS BIG AS A GOAT', PLUS A SECOND, RECENTLY-REDISCOVERED UNDERWATER CRYPTID SIGHTING AT LOCH NESS

Of all of the many Nessie-related subjects documented by me down through the years (and now collected together in this present book), few have attracted so many enquiries from readers and correspondents as the 'frog as big as a goat' supposedly sighted by diver Duncan Macdonald while underwater in Loch Ness. As I noted earlier in this chapter, the incident was first reported in Inverness's *Northern Chronicle* newspaper (by an unnamed writer) on 31 January 1934, and this report has in turn been referred to by a number of subsequent publications, but (as far as I am aware) it has never been republished anywhere in full – until now!

Courtesy of the indefatigable research skills of fellow cryptozoological investigator Richard Muirhead, in May 2016 I was delighted to receive a copy of the original *Northern Chronicle* article, which actually consisted of several different Nessie-themed items linked together. Here is the quite short but very intriguing one concerning Macdonald's alleged encounter:

A DIVER'S EXPERIENCE

Many stories have been circulated by those who go down to the depths inside a diving-bell. Some of them are, doubtless, true; others, of course, must be taken with the proverbial grain of salt. But the truth of any story can very often be guaranteed by a little careful investigation.

Here is such a story, and, as it concerns Loch Ness, and the experience which befell a well-known diver, it might, by reason of its uniqueness, act as a spur to those whose training has fitted them to probe the mysteries of marine zoology, for, in the opinion of the writer, it is but another aspect of the case of the Loch Ness "monster."

Some forty-five to fifty years ago a small sailing vessel carrying a cargo of guano, when making the passage through Loch Ness, struck a submerged reef known as "Johnnie's Point," and sank, fortunately without loss of life.

The mishap occurred during the night, and when dawn broke it was seen that the tops of the masts were still above water.

Realising that the vessel might be raised, a squad of men was quickly on the scene, and chains were passed underneath the hulk.

But ere the job was completed the action of the water suddenly dislodged the craft, and she vanished into the depths.

Still hoping to salve the wreck, the owner secured the services of Mr Duncan Macdonald, a noted diving expert, who was at the time employed at the Crinan Canal.

Mr Macdonald duly arrived, and it was from the Caledonian Canal Company's diving-barge that he carried out operations.

A STRANGE CREATURE

After having made a descent of thirty feet, Mr Macdonald signalled that he wished to come up, and, on being questioned as to whether there was any sign of the ship, he said there was none.

From this it was obvious that further attempts would be useless, so he was undressed, and the party prepared to make for Fort-Augustus, their headquarters.

Now one man in the party, having heard stories of a strange creature which was said to live in the loch, began to question the diver. The latter, however, was at first rather diffident about taking any part in the conversation.

Yet, since the others knew that anything he might tell them would be perfectly true, they persisted, and finally the diver said that he saw a strange creature that day.

It lay, he said, on a ledge of rock, on the self-same ledge, apparently, on which the keel of the wrecked vessel had rested, about thirty feet down.

There, he continued, lay a queer-looking beast, which he described as something in the nature of a huge frog.

It stared at him, but, as it showed neither ferocity nor fear, he did not disturb it. In his own words he "saw that the beast made no effort to interfere with me, and I did not interfere with it." As to size, the diver said the creature was "as big as a goat, or a good wedder [Scots dialect word for a castrated male sheep]."

The story, exactly as given, was told by Mr Donald Fraser, lock-keeper, Fort Augustus, who often heard the diver (his own grand-uncle) tell it many years ago.

STRANGE CREATURES

Naturally, this incident raises some very important questions, and the first is – Is the frog-like creature related in any way to the "monster" or "monsters" which inhabit Loch Ness?

Or does the diver's story show that such creatures are entirely different from the present "monster"?

If this be so, it is not unreasonable to presume that they might prove to be the form, or perhaps one of the forms[,] of life with which – who can tell? – Loch Ness abounds, and on which the "monster" sustains itself.

In any case[,] past reports of strange creatures having been seen in the loch show conclusively that they and their kind have had their homes there for centuries, and, this being so, it would seem that were they living on fish life, *i.e.,* salmon and trout, to the extent that some people think they do, the whole or at least most of the salmon kind – still fairly plentiful – would long since have been decimated.

Thus, there being no reason at all why the above statements should be doubted, it will surely be granted that the time is ripe for some competent body to conduct an investigation into the under-water life of Loch Ness.

This remarkable report does indeed raise some very important questions, though not necessarily the ones posed in it by its anonymous author.

First and foremost: as Loch Ness is famous for the blackness of its waters due to their high concentration of peat, how could Macdonald have perceived this goat-sized 'frog' – or indeed anything else, for that matter – while diving at a depth of 30 ft? Having said that, the very fact that he went down there at all, in search of the sunken guano vessel, suggests that some degree of underwater vision must be possible at such depths in this loch. Perhaps, however, the viewing conditions were not sufficient for him to obtain a clear picture of the creature's form, so, who knows, maybe it wasn't genuinely frog-like after all, but actually was simply a typical Nessie longneck viewed at an angle at which its neck was not visible to him.

Alternatively, there is even the possibility that in reality it was some very large form of vaguely frog-like fish – an extremely big wels catfish *Siluris glanis*, perhaps, whose wide mouth would certainly call to mind that of a frog if encountered face-on in poor visibility. The wels is not native to Britain, it was introduced to various lakes here from Germany during the 1870s and 1880s, but Loch Ness is not one of the lakes featured in documented introductions. Of course, as so many illegal introductions/releases of non-native species across Britain during the past two centuries readily testify, however, just because no documented introductions of wels specimens into Loch Ness are on record, this doesn't necessarily mean that none has taken place...

Moving on, it is well worth noting that the *Northern Chronicle*'s telling of Macdonald's sighting is very matter-of-fact, in stark contrast to modern-day retellings, which generally claim that he was terror-stricken, refused to speak about his sighting for days afterwards, etc, etc. Consequently, these would appear to be melodramatic embellishments added subsequently by person(s) unknown.

The notion aired by the above report's author that perhaps this creature was not itself Nessie but was instead some second, entirely different species of monster – and one, moreover, that may actually constitute the prey of the 'real' Nessie – offers a fascinating if implausible prospect to say the least, doubling the quandary of whether *any* type of large cryptid inhabits

Might the goat-sized 'frog' have actually been a very large wels seen by Macdonald in very poor viewing conditions? (© William M. Rebsamen)

this vast expanse of freshwater.

Equally thought-provoking is the author's claim that "...past reports of strange creatures having been seen in the loch show conclusively that they and their kind have had their homes there for centuries". On the contrary, because cryptozoological sceptics in particular habitually discount traditional stories of water-horses and water-bulls existing here, for instance, as nothing more than folk-tales, with no factual basis.

Moreover, in a very extensive *Fortean Studies* paper published in 2001 that surveyed no fewer than 87 cases of mysterious beasts allegedly spied in or on the shores of Loch Ness prior to 1933 (the year that marked the beginning the modern age of Nessie sightings), German cryptozoological investigator Ulrich Magin dismissed all of them as featuring mere legends, unsubstantiated rumours, or creatures that were unrelated to the long-necked Nessie-type cryptids reported from this loch from 1933 onwards. He concluded that there was no pre-1930s tradition of monsters inhabiting Loch Ness, only the possibility that some marine creature had somehow entered it during the early 1930s and that this is what had given rise to subsequent sightings of monsters there. This prospect is one that had been contemplated by the likes of early LNM chroniclers Lieutenant Commander Rupert T. Gould and Dr Anthonie C. Oudemans too.

But what did Magin think about Macdonald's 'giant frog'? In his listing, this was Case #32, but, interestingly, he was apparently unaware of its original *Northern Chronicle* source, because he stated: "This is a story which appears in most books about Nessie but always without reference", and he cited one such book, Peter Costello's *In Search of Lake Monsters* (1974), as the source that he had consulted.

After quoting Costello's brief version of Macdonald's own description of this underwater mystery beast's appearance, Magin concluded "...the description is unlikely to refer to a long-necked animal or any other animal known in the loch". Or indeed elsewhere, in fact, as I am certainly not aware of any living species of frog-like creature the size of a goat that is currently known to science in the living state (there are of course various extremely large amphibians known from the fossil record).

At present, therefore, the goat-sized 'frog' of Loch Ness remains a major enigma in the Nessie chronicles. Nevertheless, now that its original published source has been resurrected and reproduced here, one of the most mystifying and paradoxical LNM-associated reports – ostensibly unlikely, yet supplied by a very experienced and seemingly highly-reliable eyewitness - is finally readily available for scrutiny and further investigation by future Nessie researchers.

Finally: on 10 May 2016, shortly after I had uploaded a version of this postscript as a stand-alone article on my *ShukerNature* blog, LNM researcher Roland Watson brought to my attention a remarkable but very little-known additional underwater sighting of a mysterious beast at Loch Ness, dating from the 1880s, which had originally appeared in a newspaper-published letter of 1933, and which he had documented on his own blog, *Loch Ness Monster*, on 8 December 2013. (Magin was presumably unaware of this newspaper letter when compiling his previously-noted documentation of pre-1933 LNM cases, because he didn't include it therein.)

While checking through some old newspaper clippings in the National Library of Scotland, Watson had discovered a fascinating letter published on 9 December 1933 in the *Dundee Courier and Advertiser*. Written by someone calling themselves 'Methven', the relevant portion is as follows:

> The story runs that a gentleman sailing down the loch in his yacht was driven shorewards. The yacht sank and came to rest on a ledge of rock.
>
> In order to secure some of his valuables the owner of the sunken yacht hired the Caledonian Canal diver – Honeyman by name – from the village of Clachnaharry. He intimated to the assembled onlookers that he would begin work the following morning. When the coast was

clear, however, the diver, thinking that he and his [illegible word] would get something for themselves, donned his diving suit and descended into the water to make inspection.

When he got to the yacht it slipped off the ledge of the rock and disappeared into the depths. But then, to the diver's intense horror, a huge beast, measuring about nine feet long and possessing a body as stout as that of an average man, passed in front of him.

Perhaps the diver thought that it was the [illegible word] of the lower regions that had come to advise him on his dishonest work, or that Loch Ness possessed the door to the nether regions. Whatever it may have been, the diver, who in very truth had 'got the wind up,' adamantly refused to descend into the water again.

"I understand," added my friend, "that there is still living in Clachnaharry a man that can remember the details of this incident, which happened nearly 50 years ago. It was no unusual thing for parents of the Loch Ness district to frighten recalcitrant children by threatening to bring the Loch Ness water horse to them."

Bearing in mind that this alleged incident dated from much the same time period as Macdonald's, and featured a similar storyline too, it is tempting to speculate as to whether the two reports actually feature one and the same incident; or whether one is fictitious, having been inspired by the earlier, purportedly factual one. However, both divers were real persons (Watson has traced in the 1881 census a person who seems very likely to have been the Honeyman referred to in that report), and the localities were different.

As for the creature reputedly encountered by Honeyman: other than its size (which if estimated accurately made it bigger than anything officially known to exist in Loch Ness), there are no details of its appearance. Consequently, it is impossible to speculate with any degree of confidence what this animal may have been – a situation with which Nessie investigators will be only too familiar!

CHAPTER 8:

NESSIE AND A H⊕AX ⊕F ELEPHANTINE PR⊕P⊕RTI⊕NS!

Tomorrow, April Fools' Day, the world should know if new light has been thrown on the Loch Ness monster, or if today's find of a huge creature is just another hoax.

The carcass was found by a team from Flamingo Park Zoo, Malton, Yorkshire, about three miles south of Dores, on the east side of the loch...

On the shores of Loch Ness at Foyers I briefly examined the find. It looked like a seal, but was so enormous that all who saw it were convinced that it could be no normal seal. It measured about 12ft from nose to tail and was covered in short, thick, brownish-yellow fur. At the tail were seal-like flukes, and on the body flippers. The huge head had a mouth at least 9in wide, with sharp, pointed teeth.

'A Correspondent, Inverness' – *The Times* (London), 31 March 1972

Over the years, mystery beasts have featured in numerous April Fool spoofs perpetrated by the media. Somewhat appropriately (or even inevitably?), however, the most famous cryptozoological April Fool involved the most famous cryptozoological creature - the Loch Ness monster.

A VERY SUSPECT BODY OF EVIDENCE!
On 1 April 1972, newspapers worldwide reported the astonishing discovery of a 9-ft-long seal-like beast, weighing over half a ton, that had been found dead at the side of Loch Ness. Soon afterwards, police intercepted a van transporting the body south to Flamingo Zoo Park in North Yorkshire, England.

When examined, it was found to be the carcase of a dead bull elephant seal, which had been deep-frozen for several days, giving it a peculiar greenish hue. Its whiskers had been shaved off too, and its cheeks had been stuffed with rocks.

On 2 April, Flamingo Zoo Park's public relations officer, John Shields, confessed that it was an April Fool hoax, which he had secretly set up as a joke directed at the managing director of

A southern elephant seal (public domain)

the company owning the zoo. Shields had used the body of a southern elephant seal *Mirounga leonina* that had died a week earlier at Dudley Zoo, owned by the same company.

Another mystery solved. If only all cryptozoological cases could be resolved so readily!

Two other, more recent Nessie-linked hoaxes that hit the headlines worldwide were these:

NESSIE'S BACK...OR JUST A BIT OF IT?
It was the find that cryptozoologists had been praying for - a piece of plesiosaur backbone found in Loch Ness! Inevitably, however, things were not quite that simple. Yes, the alga-covered 12-in-long object discovered in shallow water at the loch by pensioner Gerald McSorley in July 2003 was indeed identified conclusively by palaeontologist Dr Lyall Anderson at Edinburgh's National Museum of Scotland as a series of four vertebrae from a plesiosaur - but they were fossil vertebrae, approximately 150 million years old. Could they therefore be from Nessie's far-distant loch-dwelling ancestor?

Sadly, no - if only because Loch Ness didn't come into existence until a mere 12,000 years ago, at the end of the last Ice Age. Moreover, the fossil had been intensely drilled by marine sponges, whereas the loch is freshwater; and it is embedded in a grey Jurassic-aged limestone,

FOSSIL SPECIMEN
NAME: PLESIOSAUR VERTEBRA
HORIZON: CRETACEOUS LOWER GREENSAND
LOCALITY FARINGDON, BERKSHIRE. ENGLAND
Date:

A fossil plesiosaur vertebra (© Dr Karl Shuker)

yet rocks in the Loch Ness area are much older and are all crystalline, igneous, and metamorphic. According to Anderson, the nearest match for the fossil's limestone is some 30 miles away, at Eathie on the Black Isle.

Consequently, it seemed much more plausible that the fossil had simply been deliberately planted at Loch Ness as a hoax, the latest in a long line associated with Nessie over the years, in the hope that someone would eventually come along and find it, as indeed happened.

Conversely, plesiosaur expert Dr Richard Forrest from Leicester's New Walk Museum was quoted in media accounts as suggesting that the fossil may have reached Loch Ness by more natural means - transported by glaciers.

There is also a telling argument against the hoax theory that I have not seen mentioned in print before, but which immediately occurred to me after reading about this incident: namely, plesiosaur fossils of this type are neither cheap nor commonplace. As someone who regularly peruses mineral and fossil fairs, I have seen single plesiosaur vertebrae on sale for around £40 (the sum that I myself paid for one such specimen, pictured on p.127). Hence I would expect a four-vertebra example (should such a specimen actually be available to buy) to cost in excess of £100 - which therefore seems an expensive item to discard purposefully in Loch Ness just on the mere off-chance that it would one day be found.

NESSIE'S REAL – AND THAT'S THE TOOTH!...OR IS IT?
In May 2005, media worldwide carried reports of how, just a couple of months earlier, two unnamed American law students visiting Loch Ness had allegedly discovered a very sharp, 4-inch-long, tooth-like object embedded in the ribs of a deer carcase on the shores of the loch. They then flagged down a passing boat, but this was supposedly manned by a Scottish governmental official who confiscated the tooth and also a videotape that the students had filmed of it.

Fortunately, however, they also had some footage of this object on a second, undisclosed videotape, which they subsequently showed to an American researcher of the paranormal named Bill McDonald. He in turn appeared on a Canadian radio show whose host, Rob McConnell, subjected him to a very meticulous inquisition concerning his Nessie-linked claims regarding the object. By the end of the show, McConnell was convinced that the whole story was a hoax, and after photos of the so-called tooth had been closely inspected by several British and Canadian zoologists he released a press statement stating that it wasn't a tooth at all, but merely a prong from an antler of a muntjac deer.

It eventually transpired that the truth about the 'tooth' was that the entire incident was a publicity stunt to promote a new horror novel by bestselling author Steve Alten (for whom McDonald was working) entitled *The Loch*.

The official but anonymous webpage devoted to the 'tooth' and its potential scientific significance, created prior to its exposure as a deer prong, can still be viewed online, at http://www.lochnesstooth.com/

FINDING NESSIE!...SORT OF
Finally: When in mid-April 2016, media reports announced that Nessie had been found, even the most optimistic cryptozoologists suspected that such a statement was too good to be true – and indeed it was. Happily, however, the reality on this occasion was not simply another much-expected hoax, bur rather a very unexpected rediscovery.

During April 2016, a state-of-the-art, torpedo-shaped, underwater robot named Munin, operated by a team from Norwegian company Konigsberg Maritime (and led by Craig Wallace) in co-operation with Adrian Shine of The Loch Ness & Morar Project, and VisitScotland, conducted a very extensive two-week survey of Loch Ness. Sadly, however, although Munin found no trace of Nessie herself, it did indeed find a monster of sorts there, resting placidly and undisturbed upon the loch floor where it has been for nearly 50 years now, hidden from sight but never forgotten.

At a depth of over 700 ft, thereby confirming its technical expertise in underwater exploration, Munin had obtained sonar scans of a very large object whose size, shape, and location are all entirely consistent with the huge, long-lost Loch Ness monster model that had been specially constructed for the 1970 British movie *The Private Life of Sherlock Holmes*, starring Robert Stephens as the great detective plus Christopher Lee as his brother Mycroft, and directed by the legendary American film-maker Billy Wilder. Unfortunately, however, the movie's spectacular plesiosaur-like Nessie model, approximately 30 ft long, and complete with a tall vertical neck and head, had abruptly sunk to the bottom of the loch when it was towed out onto the water ready for filming in 1969 during the making of the movie, after Wilder had ill-advisedly requested that its two humps be removed (which apparently disrupted its balance, causing it to overturn and sink).

This amazing prop had been designed and built by special-effects wizard Wally Veevers, who would later garner much-deserved praise and awards for his celebrated work on such classic films as *2001 A Space Odyssey*, *Superman*, and *Local Hero*, and he was so upset when his Nessie model sank that he had to be personally comforted by Wilder. A new Loch Ness monster model was constructed afterwards as a replacement, but this time only a head and neck were built, and filming took place not on the loch but in a large studio-based water tank instead, just to be on the safe side.

After all, as Oscar Wilde's Lady Bracknell might have observed had she been there: "To lose one Nessie may be regarded as a misfortune; to lose both looks like carelessness".

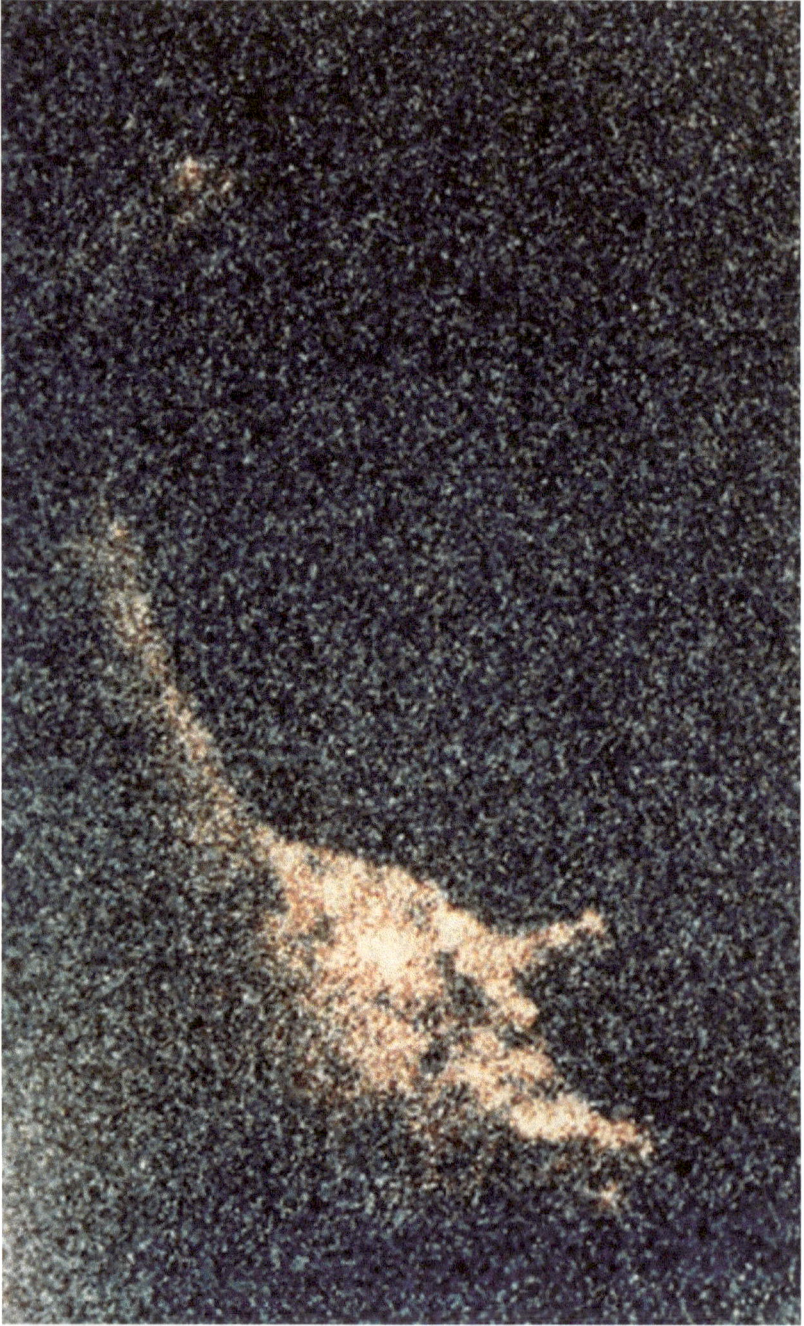

Nessie, is this really you? The 'head and neck' underwater photograph snapped at Loch Ness in 1975 by Dr Robert H. Rines's Academy of Applied Science LNM research team (© Dr Robert H. Rines/ Academy of Applied Science)

THE MANY FACES OF NESSIE
- A LOCH NESS MONSTER PICTURE
GALLERY IN FULL COLOUR

Over the years, I have been fortunate enough to become friends with many extremely gifted artists, including a number who specialise in cryptozoology-themed artwork, including the Loch Ness monster. I have also collected and received many photographs relating to the LNM, its possible identities, and localities associated with it.

So I now have great pleasure in presenting an exclusive LNM picture gallery that includes a delightful series of Nessie depictions, representations, and portraits deftly combining individual expression with zoological verisimilitude, plus a wide selection of pertinent photographic images - all of which are in full colour and collectively yield a diverse and very dramatic array of spectacular LNM-themed illustrations.

Many of these pictures have never been published before, having been prepared or obtained specifically for inclusion within this present book. So to everyone who very kindly volunteered their time and considerable talent to this project, either in producing artwork or in assisting me with the procurement of photographs, please accept my most sincere thanks!

The Surgeon's Photograph – it may be b/w, but not even a Nessie gallery devoted to full-colour pictures could justify not including the world's most iconic and controversial LNM-related image, so here it is! (Fortean Picture Library)

Monsters By Moonlight (© Wm Michael Mott)

Nessie as a long-necked seal (© Anthony Wallis)

Loch Ness (public domain)

Here's Nessie! (© Dr Karl Shuker)

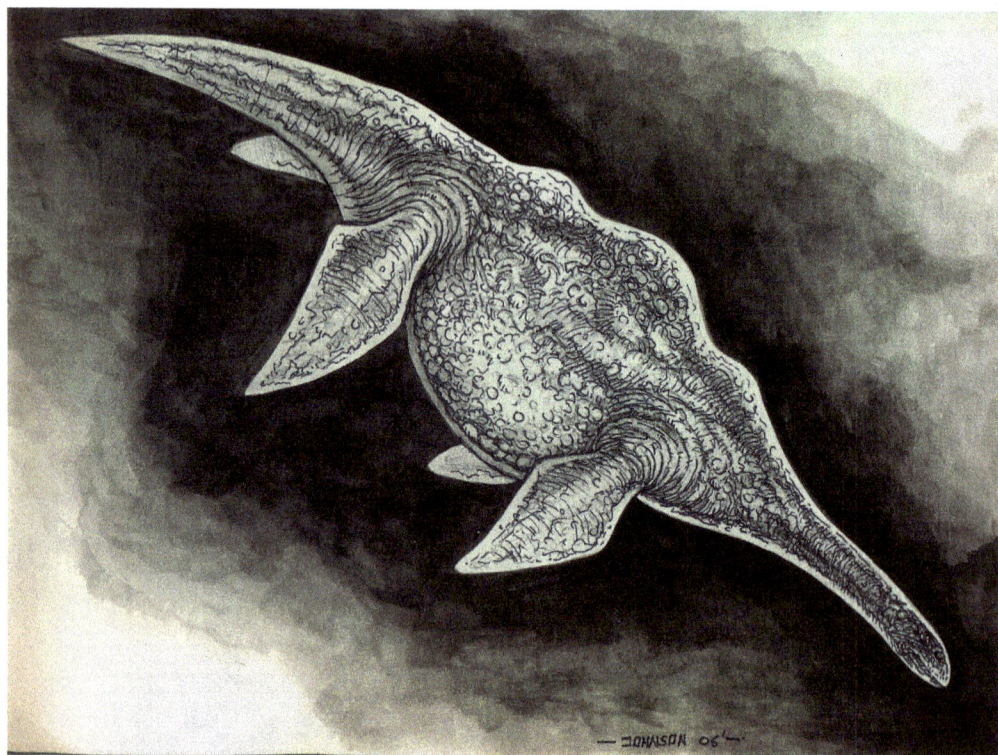

A plesiosaurian Nessie inspired by the 'flipper' photographs from 1972
(© Jeff Johnson)

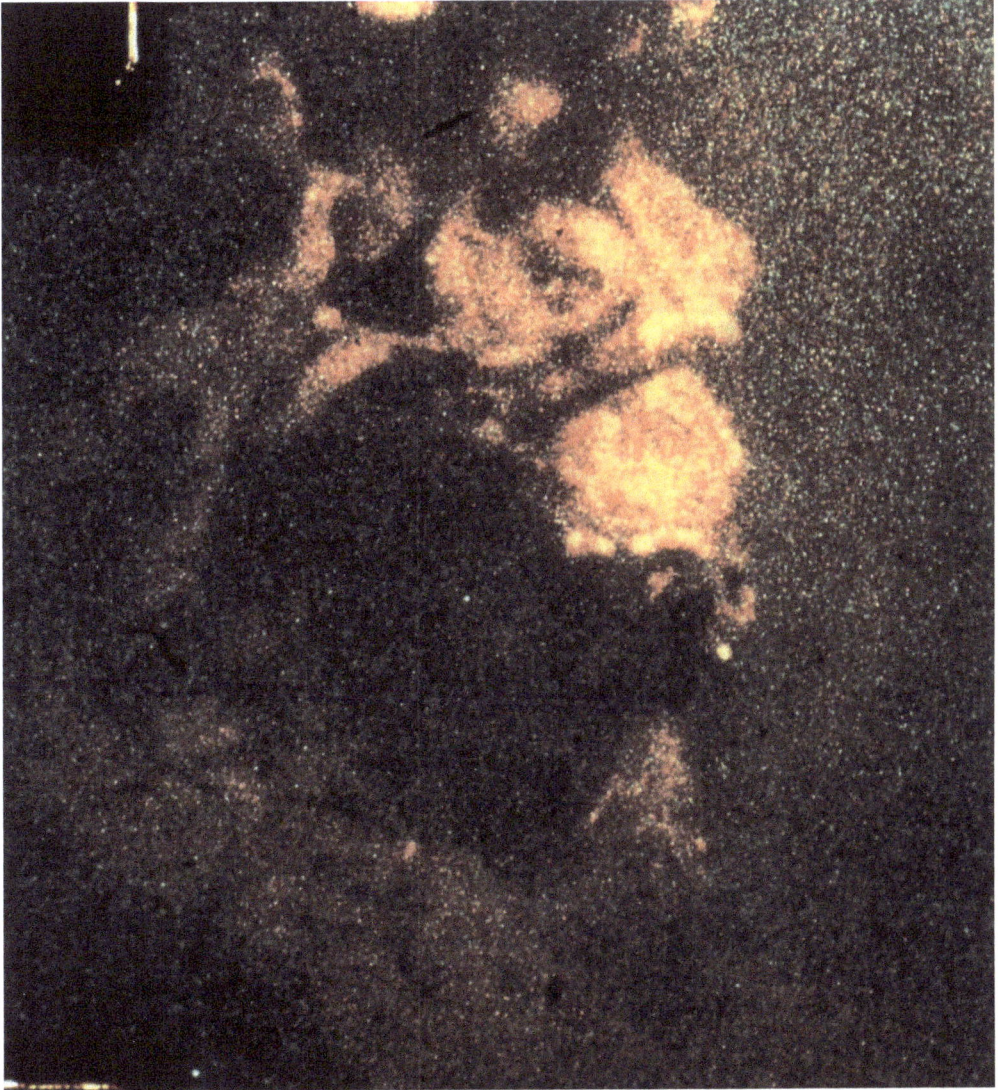

The 'gargoyle' photograph (© Dr Robert H. Rines/Academy of Applied Science)

A modern-day Nessie as *Megalotaria* against a backdrop of James Parsons's long-necked seal illustration from 1751 (© Robert Elsmore)

A family of Nessies (© Richard Svensson)

Life-sized plesiosaur model on land (public domain)

Swan-necked seals *Acrophoca longirostris* depicted in a mural at the Staatliches Museum für Naturkunde Karlsruhe in Germany (© Markus Bühler)

An eel adopting a Nessie-like 'periscope' pose underwater
(© Dr Karl Shuker)

Even normal otters can yield Nessie-like outlines when observed in certain lights and at certain angles in water (© Dr Karl Shuker)

Could Nessie be a giant form of aquatic salamander? (public domain)

Tony 'Doc' Shiels's painting of his hypothetical elephant squid
(© Tony 'Doc' Shiels)

Pictish beast depicted on Rodney's Pictish Stone at Brodie (© Ann Harrison/
Wikipedia http://www.creativecommons.org/licenses/by-sa/2.0/)

Dobhar-chú drawn to scale against a human figure
(© Connor Lachmanec, aka TheMorlock/deviantart)

Kelpie (© Dr Karl Shuker)

Water-bull (© Steve Garvie/Wikipedia http://creativecommons.org/licenses/by-sa/2.0/, adapted by Dr Karl Shuker)

'Home of the Waterhorse' sign at the Loch Ness Monster Exhibition Centre, Loch Ness (public domain)

A water-horse's fiery eyes always betray its true identity (© Dr Karl Shuker)

The lighter side of Nessie (© Jane Cooper)

A humorous take on the long-necked seal concept (© William M. Rebsamen)

Sheet music to the *El Plesiosaurio* tango (public domain)

Nessum-monstrosum – humorous Nessie illustration in sepia from 1934 (public domain)

A three-dimensional model of the LNM (© Jeff Johnson)

Nessie as an elasmosaurid plesiosaur (© Peter Loh)

A plesiosaurian Nessie emerging at Drumnadrochit (public domain)

Is this how Nessie might look if seen on land at dusk? (public domain)

Is Nessie a plesi? (© Dr Karl Shuker)

Is Nessie an antediluvian amphibian like *Eryops*? (© Dr Karl Shuker)

Is Nessie otterly mysterious? (© Dr Karl Shuker)

Are crocodiles a snappy solution to Welsh Nessies? (© Dr Karl Shuker)

Welcome to my underwater world – I've been expecting you!
(© William M. Rebsamen)

Serpentiform reconstruction of plesiosaurs, from *Extinct Monsters - A Popular Account of Some of the Larger Forms of Ancient Animal Life*, 4th ed (1896), by the Reverend Henry N. Hutchinson
(public domain)

The first Nessie book, published in 1934 (public domain)

Two LNM centres at Loch Ness (public domain)

Loch Ness, vintage sepia postcard (public domain)

Loch Ness, The Great Glen at Urquhart Bay, vintage sepia postcard
(public domain)

Castle Urquhart and Loch Ness, vintage pre-1914 picture postcard (public domain)

Loch Lomond Head, looking north, vintage picture postcard (public domain)

A small selection from my LNM book, video, and DVD collection
(© Dr Karl Shuker)

My collection of Nessie and other water monster figurines (© Dr Karl Shuker)

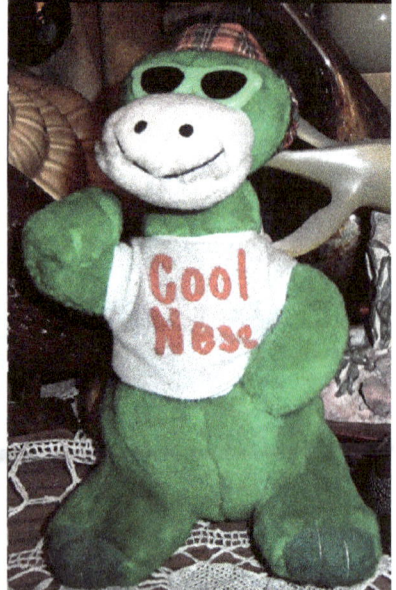

**Some Nessie-inspired fluffy toys from my collection of LNM memorabilia
(© Dr Karl Shuker)**

I had Nessie all along! (© Dr Karl Shuker)

Loch Ness, painted by Alfred de Bréanski (1852-1928) (public domain)

Loch Ness, painted by a follower of Samuel Bough in c.1877 (public domain)

The classic Nessie look? Modern restoration of the likely appearance in life of the Cretaceous elasmosaurid plesiosaur *Styxosaurus* (© Julian Johnson CC BY-SA 2.0 licence http://creativecommons.org/licenses/by-sa/2.0/legalcode)

Peek-a-boo - I see you! (© Julian Johnson CC BY-SA 2.0 licence http:// creativecommons.org/licenses/by-sa/2.0/legalcode)

Nessie envisaged as a highly-evolved long-necked seal (© Robert Elsmore)

LEFT: 'The Fairy Queen Lures the Water Dragon' (© Thomas Finley)
RIGHT: Tim Dinsdale, a personal tribute by Thomas Finley (© Thomas Finley)

Chapter 9:

A COUPLE OF NESSIE-SSARY REVIEWS

In discussing these exceptions from the course of nature, the first question is, whether the fact be justly stated. That which is strange is delightful, and a pleasing error is not willingly detected. Accuracy of narration is not very common, and there are few so rigidly philosophical, as not to represent as perpetual, what is only frequent, or as constant, what is really casual...Natural philosophy is now one of the favourite studies of the Scottish nation, and Lough Ness well deserves to be diligently examined.

Samuel Johnson – 'Lough Ness', in *A Journey to the Western Islands of Scotland*

Over the years, I have reviewed numerous books and other media releases appertaining to cryptozoological subjects, including the following two, both of which focus significantly upon Nessie.

LOCH NESS DISCOVERED
- REVIEWING THE DVD.

Discovery Channel, 2005, approximately 1 hr 17 min.

The cover of this Discovery Channel DVD prominently displays the familiar image of the Surgeon's Photograph, purportedly depicting Nessie, but as its title suggests, the scope of the DVD's contents goes beyond Nessie to encompass Loch Ness as a whole. Indeed, of the four films included on it, the principal one, sharing its title with the DVD itself and lasting for 45 minutes, is primarily concerned with the loch's natural - as opposed to unnatural - history.

Originally released in 1993, Film #1 follows Project Urquhart, featuring the researches of two scientific teams working at Loch Ness, studying its complex but hitherto little-investigated underwater ecosystem. One team, from the Freshwater Biological

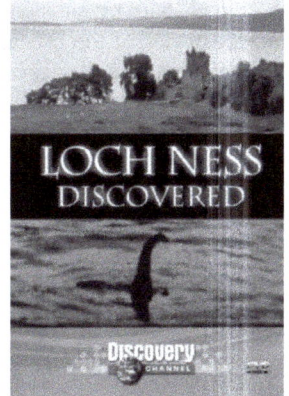

Association (FBA), is particularly interested in the intriguing fact that Loch Ness is to all intent and purpose two separate lakes, comprising a warmer upper layer, where little lives, above a colder, wilder, under layer containing fauna and dramatic underwater weather. The second team, from London's Natural History Museum, is surveying global pollution, which it is investigating at Loch Ness by examining its microscopic but pollution-sensitive nematode worms.

Interspersed with coverage of these ongoing mainstream studies are cryptozoologically-interesting segments focusing upon various aspects of the Loch Ness monster phenomenon - such as Peter MacNab's 1955 photo of a Nessie-type form close to Urquhart Castle, the underwater 'flipper' photos of Dr Robert Rines, the Surgeon's Photograph, Tim Dinsdale's film, assorted eyewitness accounts, and psychologist Dr Susan Blackmore's theories of what may be influencing such accounts. Along the way, some intriguing data and findings emerge.

For example, in the past, sonar has found a series of strange regular prints on the loch bed, nicknamed 'the footprints', whose origin has never been explained, but which may be related to wartime military exercises here. During the two teams' studies, a remote-controlled unmanned craft, the *Sea Owl*, filled with cameras, is sent down to investigate one of these prints, but reveals it to be nothing more startling than a submerged wheelbarrow. Surely, however, as aptly queried by the narrator, submerged wheelbarrows couldn't explain all of these 'footprints', but then he seems to run out of investigative steam, ending with the weak comment that scientists can only speculate. Why can they only speculate? Bearing in mind that the *Sea Owl* had successfully unveiled the identity of one of these prints, how little more time, trouble, and money would it have required simply to have taken this craft along the loch bed a bit further while it was already there, in order to spy on a few more of these prints and find out what they were too? Surely this was a superb opportunity to solve at least one Loch Ness mystery that instead was needlessly lost.

A very notable, unexpected find made by the Natural History Museum team's fish expert, Dr Colin Bean, was that, contrary to a previous estimate, in 1973, that the loch contained 3 tons of fish (and which had been deemed sufficient to support a higher predator), it now appears that a much more realistic estimate is 27 tons. That is, 9 times more fish than hitherto assumed, thereby substantially increasing the possibility that the loch could sustain a large-sized species of top predator - a loch which, incidentally, contains as much water as in all of England and Wales's lakes combined.

Following this discovery, the film proceeds to consider the biology of plesiosaurs, deemed the best fit for most Nessie sightings, as well as for Rines's flipper photos. Using the computer enhancement expertise of Brian Reece Scientific Ltd with the original unenhanced photos, the researchers attempt to duplicate the final rhomboid flipper images widely publicised by the Rines team, but are unable to do so. Moreover, when they examine the surgeon's photo, they notice a curious white spot just in front of the neck, which may indicate the presence of something towing the neck along, but equally may just be a blemish on the negative.

The most interesting find made when applying computer enhancement expertise, however, occurs with a frame depicting a very large object moving across the loch from the famous film shot by Tim Dinsdale in 1960. First of all, the team examines not just the frame's positive but also its negative image, and is surprised to see in the negative a shadow behind the object. Furthermore, when the positive is cleaned up by enhancement techniques, a very large underwater shadow directly beneath the object can clearly be seen - implying that whatever this object is, it possesses an extremely sizeable hitherto-unsuspected portion present beneath the water surface, and thereby arguing against the possibility that it is merely a surface vessel such as a boat. (Having said that, Nessie investigator Dick Raynor has opined at http://www.lochnessinvestigation.org/cyberspace.html#seeunderwater that perhaps the underwater shadow is merely a reflection of the shore behind the object, but this seems highly unlikely to my mind.)

The climax of the film, however, comes with the FBA's sonar work aboard their research vessel *Calanus*. During the evening of 19 July 1993, a massive underwater storm is recorded by their sonar equipment as it rages beneath the vessel, an event rarely witnessed before, and guaranteed to disturb the loch's fauna. The following day, while examining the sonar traces recorded during that storm, Dr Colin Bean and other members spot a very large, unidentified sonar trace deep in the water with a second one close by (and perhaps even a third and fourth), which do not appear to be shoals of fish because they are followed by quite a pronounced wake (whereas shoals of fish do not cause wakes). The team members are perplexed, unable to explain these anomalous traces.

The film ends in celebration - what appears to be a totally new species of microscopic nematode worm has been discovered during the research work. Cryptozoologists, however, may wish that the unexplained sonar traces had elicited as much interest and attention.

Film #2, entitled *PaleoWorld: The Loch Ness Secret*, and lasting 25 minutes, is probably of more direct cryptozoological pertinence, as it attempts to uncover the possible identity of Nessie, by examining three supposed contenders from prehistory - ichthyosaurs, mosasaurs, and plesiosaurs. British palaeontologist Dr Michael Benton discusses the anatomy and lifestyle of each one, supplemented by various specialists from elsewhere around the world and some stunning film of preserved fossils, as well as a reconstruction of pioneering fossil hunter Mary Anning's discovery at Lyme Regis, Dorset, during the 1800s of the first complete ichthyosaur and plesiosaur skeletons.

Personally, I found the ichthyosaur segment superfluous, as this remarkably fish-like or even dolphin-like reptile bore little if any resemblance to eyewitness accounts of Nessie. Indeed, the most memorable part of it came at the very end, with the narrator's chilling closing line - noting that if ichthyosaurs do indeed exist in Loch Ness, it could be the most dangerous place in the world to go fishing! Other than Lake Champlain, perhaps?

Sandwiched between the ichthyosaur and mosasaur segments is a reconstruction of the mystifying land sighting by chauffeur Alfred Cruickshank, which occurred at dusk one evening in summer 1934 according to this film (but normally given by other sources as early morning in April 1933) as he was driving along the north bank of the loch. At the crest of a hill, his car's headlights picked out a big animal crossing the road. It had a large humped body, estimated at 4 ft high and around 25 ft long, and waddled away on two pairs of legs, its belly on the ground, and its head close to its body, with very little neck. Later, summing up the mosasaur section, the novel question is posed as to whether Cruickshank's mystery beast was a female mosasaur that had come on land to lay her eggs and was now returning to the loch, just as sea turtles come ashore to lay their eggs before going back into the sea.

The third, and most popular, reptilian contender for Nessie is then discussed - the plesiosaur. Included here is an eyewitness reconstruction from 1 June 1994, when, after seeing a mysterious object above the water surface while driving alongside the loch, Fiona Mackay and her friend Errol David jumped out of their car and ran along the bank for a clearer view. The object had a long tall neck and moved swiftly in the water, then suddenly dived, creating such a splash that its two observers had to jump back to avoid being soaked. Moreover, other eyewitnesses saw it that night. However, the film ends with no firm suggestions as to what Nessie may be, always assuming that such a creature does exist.

Films #3 and #4 are no more than a few minutes long. The first of these is a brief interview with Adrian Shine at the onset of Operation Deepscan back in 1987; and the second, less than 2 minutes long, is a montage of film clips of early Nessie expeditions, and images as to what it may look like.

All in all, this DVD is an interesting survey not just of Loch Ness as a famous 'monster' lake, but also as a body of water that is actually as puzzling to mainstream zoology as it is to cryptozoology (though it should be borne in mind that as these films were made during the 1990s, their findings are not current). If you are hoping for an exclusively cryptozoological package, you may be disappointed, but worthy of note here is that the cryptozoological coverage is presented in a relatively optimistic, open-minded manner - in stark contrast to the depressing tendency by so many of the more recent LNM documentaries to rule out of hand with smug self-assurance even the faintest possibility of a cryptozoological mystery existing here.

LOCH NESS, NESSIE & ME
by Tony Harmsworth.

Harmsworth.net: Drumnadrochit, 2010. Pb, 356 pp, b/w illustrations, index, photograph list, references. ISBN 9781456380236

Over the years, virtually every major personality in the long-running cryptozoological saga of the Loch Ness Monster (LNM) has published one or more books, reports, or scientific papers on this ever-intriguing subject – Prof. Henry H. Bauer, Dr Maurice Burton, Tim Dinsdale, F.W. Holiday, Prof. Roy Mackal, Dr Robert Rines, Sir Peter Scott, Adrian Shine, Nicholas Witchell, even notorious photo-hoaxer Frank Searle. Indeed, only two

A small selection of my Nessie-related books, videos, and DVDs, including my copy
of the *Loch Ness Discovered* DVD and Tony Harmsworth's book
Loch Ness, Nessie & Me (© Dr Karl Shuker)

notable omissions from this eminent list of Nessie-linked names come to mind. One is Steve Feltham, who has spent much of the past 20 years living in a converted mobile library on the shores of Loch Ness, hoping to make that breakthrough Nessie sighting one day, and who will assuredly write a book of his extraordinary life at some point in the future. The other is Tony Harmsworth, founder of the original LNM exhibition centre in the Scottish Highlands village of Drumnadrochit, on the west shore of Loch Ness, during the early 1980s, but whose long-awaited story has now finally been published – and what a story it is.

Unlike previous Nessie books, this volume is not devoted exclusively to the LNM. Instead, it is Harmsworth's own autobiography, documenting an eventful life, but in which Nessie has certainly played a major part. Consequently, its text, arranged chronologically as one would expect with a book of this nature, has for the most part a much less formal style than those of its predecessors. Only when dealing with various key pieces of evidence, such as the sonar traces and controversial 'flipper' photos, does it become rather more technical. Inevitably, it covers much the same ground as other Nessie books when documenting the LNM's history, but what makes it unique and particularly interesting is its personality-driven format, so very different from the vehemently objective style of presentation typifying previous LNM coverages.

As his life story unfolds, Harmsworth reveals all manner of Nessie-linked facts and insights that have not previously been documented or widely publicised. His prolonged but ultimately unsuccessful battle to retain part-ownership of the exhibition that he conceived is especially revealing. So too is the saga of 'Nessie Hunt', an award-winning LNM-seeking board game that he also conceived, and which received considerable critical acclaim Sadly, however, it never attracted a comparable degree of commercial success, despite receiving some welcome publicity from none other than Doctor Who – or at least the actor Colin Baker who played the television Time Lord during the mid-1980s. Another hitherto-unsuspected event is the somewhat hapless albeit unfortunate loss en route by aeroplane from the States to Scotland of an 8-ft-long fibre-glass replica of one of the famous flipper-like images snapped underwater at the loch by Rines's research team during the early 1970s. Imagine having to claim for 'One Loch Ness Monster flipper' on the Lost Property form at the airport!

What is most entertaining and informative of all, however, is Harmsworth's own take on every noteworthy LNM personality (having met them all at one time or another during his extensive, ongoing involvement with this most famous of all cryptozoological cases), and also on every newsworthy piece of research or evidence brought to public attention during the previous decades, including all of the most celebrated photos purported at one time or another to show Nessie. He is not afraid to state his own views, and his gradual transformation from an enthusiastic optimist to a somewhat-resigned apparent agnostic (notwithstanding his own putative Nessie sighting) as to the existence in the loch of a bona fide mystery beast makes absorbing reading. So too does his elucidation of how Nessie herself has metamorphosed in the eyes of her eyewitnesses through that same period, from a large fish to a Jurassic Park-style prehistoric survivor of plesiosaurian persuasion and then back again to a large fish – with such metamorphoses seemingly effected more by outside influences than by what the eyewitnesses were actually observing.

The book is illustrated by more than 200 b/w photographs, but although an entertaining read, it would certainly have benefited from an experienced proof-reader, because a fair few typographical errors have crept in. So too, rather more worryingly, have some factual mistakes of the kind that should have been readily spotted and eliminated. For instance, when documenting a certain extremely famous bestselling author and Conservative Party peer, Harmsworth consistently refers to him not as Jeffrey Archer but as Geoffrey Archer (who is a totally separate writer of thriller novels). Similarly, on p. 142, "MP Charles Brandreth" should read "future MP Gyles Brandreth" (Charles was his non-MP father, and Gyles himself was not an MP during the period documented on that page). And when ruminating upon the annual Eurovision Song Contest, Harmsworth states "It is always thoroughly enjoyable" (i.e. present tense) listening to UK compere Terry Wogan's much-loved humorous asides – even though Wogan had actually resigned from this post back in 2008 (having been replaced from 2009 onwards by Graham Norton). One can only hope, therefore, that his Nessie coverage, a subject upon which he is no doubt far more knowledgeable, does not contain any such slips.

That aside, however, I certainly recommend Harmsworth's book as an enjoyable, often eye-opening, and very personal view of the LNM's colourful history and enduring mystery, and I for one will in any event always be thankful to him for providing one of the most memorable cryptozoological statistics ever – which I freely confess to having used on more than one occasion myself since. While chatting as a guest on 'Saturday Superstore' (a British teenage television show) one autumn morning in 1985, Harmsworth stated: "You can put the entire population of the world, every man, woman, and child on Earth, in Loch Ness THREE times over. And you'd still have room for a few mysteries." Surely, then, somewhere amid that unutterably vast volume, there is enough room for at least a few monsters too...isn't there?

(This second review of mine is reprinted from the *Journal of Scientific Exploration*, vol. 25, no. 2, pp. 404-406 (2011), published by the Society for Scientific Exploration, http://www.scientificexploration.org)

Sheet music for 'I'm the Monster of Loch Ness', a 1934 song made famous by British variety star Leslie Holmes (public domain/supplied by Joe Mancini)

Chapter 10:
In Tune With Nessie
‑ A Selection of LNM‑Themed Songs

I'm the monster of Loch Ness.
Och aye, och aye, oh yes.
Billions of barnacles are clinging to my tail.
Every night is my delight to gobble up a whale.
I'm the monster of Loch Ness.
That's my permanent address.
Though my form was never built
For a sporran or a kilt,
I'm the monster of Loch Ness.

Ralph Butler and Will E. Haines - '
I'm the Monster of Loch Ness'

It will probably come as no surprise to discover that such an iconic figure as Nessie has been celebrated and immortalised by music down through the decades since her modern-day media debut during the early 1930s, but what may be surprising is the wide range of genres that have done so - from foxtrots and folk to heavy metal, skinhead reggae, and cartoon classics. So here is an annotated listing of some famous and not-so-famous musical tributes to the world's favourite monster, whatever your tuneful tastes may be, and accompanied wherever possible with links to their performances on YouTube.

And where better to begin than with some wonderful recordings inspired by and released during that fateful 1933-1934 period of LNM-related frenzy, a period that witnessed the reporting of some classic Nessie sightings following the opening in 1933 of a new motoring road, the A82, directly overlooking the northern shoreline of this hitherto-secluded loch – a significant event that brought the alleged existence of Nessie to the attention of an entranced media, both nationally and internationally.

Leslie Holmes (1934). 'I'm the Monster of Loch Ness'.
Perhaps the most popular of these early 1930s Nessie recordings is this delightful ditty, a

comedy foxtrot written and composed by Ralph Butler and Will E. Haines, and most famously sung by British variety star Leslie 'the smiling vocalist' Holmes. Recorded by him on 6 January 1934 as a 78 rpm shellac record on London's Regal Zonophone label (with sheet music published by Cameo Music), it also featured the Midnight Minstrels, plus Scott Wood and His Orchestra. Holmes was also filmed singing it, in b/w, by London's British Pathé Studio, in an amusing sketch that included an appearance by Nessie herself at its close. Recorded on 25 January 1934, this sketch can be viewed here on YouTube:
https://www.youtube.com/watch?v=_G5NArDuK2M

In 2014, M. Ryan Taylor brought out a book of spooky songs entitled *The Haunted Ukelele*, which included 'I'm the Monster of Loch Ness'. Here is a recording of him singing it while playing a Koa model Godin Multiuke:
https://www.youtube.com/watch?v=cOcL7FvMnjo

My sincere thanks to Facebook friend Joe Mancini for first alerting me to the Leslie Holmes version of this song.

Brian Lawrance (1934). 'Boo, Boo. Here Comes the Loch Ness Monster'.
This song is much more obscure than the previous one, despite featuring the well-regarded British vocalist and band leader Brian Lawrance on its best known version, which again was recorded in January 1934 as a 78 rpm record, but this time by the Eclipse label. As yet, I have been unable to trace an online version of it.

John Tilley (1933). 'The Loch Ness Monster'.
Not a song as such but what was back then a very famous comedy monologue for radio, spoken by John Tilley, a briefly popular, quintessentially English broadcaster/revue artist during the early 1930s, who recorded it as a 78 rpm record in December 1933 for the Columbia label. Tragically, Tilley was only in his late 30s when he died in 1935.
https://www.youtube.com/watch?v=7zTVgcvJlOk

After those early recordings, a fair few years went by before Nessie received much in the way of further musical mileage, but from the 1960s (and especially the 1970s) onwards, she has been a perennially popular subject for songs and melody, as the following diverse selection demonstrates.

Robin Hall & Jimmie MacGregor (1961). 'The Monster of Loch Ness'.
This Scottish folk duo formed in 1960 and recorded over 20 albums together before their partnership ended in 1981. Their humorous Nessie song was co-written by MacGregor and was released as a 45 rpm vinyl single in 1961 on the Decca label.
https://www.youtube.com/watch?v=8sq-T-3aTWs

King Horror (1969). 'Loch Ness Monster'.
This highly-collectible 1969 single by skinhead reggae act King Horror (originally a calypso singer, apparently) and issued on the Grape label is (in)famous for the OTT bloodcurdling screams at the onset. Somehow, I don't think that his Nessie is the shy, retiring, piscivorous type!
https://www.youtube.com/watch?v=w4iXREusIS0

Alex Harvey (1977). 'Alex Harvey Presents: The Loch Ness Monster'
Best known as the founder and frontman of the Sensational Alex Harvey Band (SAHB), Scottish rocker Alex Harvey also recorded this 40-minute spoken documentary solo album, released in 1977 as an LP and now out on CD..

Harvey had previously spent the summer at Invermoriston in the Scottish Highlands by himself while the rest of his band were doing other things, and had interviewed a range of LNM eyewitnesses and historians, recording their narratives and interspersing them with views of his own and also additional commentary by Richard O'Brien of *The Rocky Horror Show* and *The Crystal Maze* fame. Released as a limited edition album by the K-Tel label and complete with an illustrated 16-page diary-format booklet in a gatefold sleeve, allegedly only around 300 copies were actually pressed, thus making it highly sought-after. It only contains one (very short) song, right at the end of the LP, entitled 'I Like Monsters Too':

https://www.youtube.com/watch?v=HSm60wG1KHQ

The Sensational Alex Harvey Band (1978). 'Water Beastie'.
The SAHB with Harvey fronting also recorded a Nessie song, 'Water Beastie', which appeared as track #8 on their 1978 album 'Rock Drill', and was co-written by Harvey, Chris Glen, and Hugh McKenna, all from SAHB.
https://www.youtube.com/watch?v=D01sdGCfRtc

The Police (1983). 'Synchronicity II'.
Appearing as track #6 on this seminal English rock band's album 'Synchronicity' and also released as a single in 1983, this song tells of two unrelated events that are happening simultaneously – a demeaned, harried husband and father's life descending into increasing depression and despair, while, far away, a monstrous entity emerges from a dark Scottish loch and moves ominously, inexorably, towards a lochside cottage. It was written by the band's lead singer/bassist, Sting.
https://www.youtube.com/watch?v=o5FPPoLqkCk

One of the most popular animated children's TV series in Britain during the early 1980s was

The Family Ness, which was created by English cartoonist Peter Maddocks of Maddocks Cartoon Productions, consisted of 25 five-minute episodes, and was originally screened on BBC 1, beginning in 1983. As its name suggests, its stars were a family of Nessies, plus two children, Elspeth and Angus MacTout, who could call the Nessies from their loch using secret thistle whistles. Each of the Nessies (of which there were many) was punningly named after their defining trait, and included among their number Clever Ness, Grumpy Ness, Lovely Ness, Hungry Ness, Silly Ness, and the daunting Ferocious Ness. The opening titles of each episode were accompanied by a catchy song, and a second equally catchy song accompanied the end credits of each one. Both were written by English songwriter Roger Greenaway and music composer Gavin Greenaway (Roger's son), and in 1985 they were released by the BBC in single format:

The Family Ness (1983). 'The Family Ness'.
This is the song played over the opening titles to each episode of *The Family Ness*. No vocalist screen credit was aired, but it has been suggested that Gavin Greenaway himself may have been the singer, as he was aged approximately 20 at that time, and the singer sounds like someone of around that same age. (ditto for the end credits song too – see next song entry).
https://www.youtube.com/watch?v=oFR0brxrxjI

The Family Ness (1983). 'You'll Never Find a Nessie in the Zoo'.
An extended, full-length version of this song, hitherto played only in brief, incomplete form over each episode's end credits, appeared in the very last episode of this TV series, and became the video for the song when released as a single. My dear little Mom, Mary Shuker, absolutely adored both the song and the video, and whenever I played it (having taped it on videocassette) she would always stop whatever she was doing and watch it, laughing with delight. Happy days, happy memories.
https://www.youtube.com/watch?v=27oxkSYFFbg

Stuart Anderson (1992). 'Nessie (The Loch Ness Monster)'.
Mom was also a big fan of this act. In 1989, at the tender age of six, pint-sized Scottish singer Stuart Anderson's highly-polished performance of 'Bonnie Wee Jeannie McColl' in the annual 'Young Entertainer of the Year' competition staged on BBC1's Saturday morning teenage television show *Going Live* so entranced the voting public that he ultimately won it by the biggest margin of votes ever recorded in this competition. On the back of his success, Stuart went on to release several albums, one of which, 'Stuart Anderson Acts Naturally', released in 1992, contained the cheery singalong song 'Nessie (The Loch Ness Monster)'. Today, aged 32 and a well-respected guitar teacher, Stuart's very youthful showbiz days are long behind him, but he remains forever young - and forever singing about Nessie (not to mention Bonnie Wee Jeannie McColl!) - on YouTube:
https://www.youtube.com/watch?v=IOiKySTzhGI

Phyllis Logan (1992). 'Shy Girl'.
This song comes from a British animated feature film entitled *Freddie as F.R.O.7.*, which was released in 1992 (and retitled as *Freddie the Frog* in the USA). A James Bond parody, it tells the somewhat complicated story of Frederic, a medieval prince and heir to his country's

throne, who is turned into a frog by his evil aunt Messina (who has already secretly killed his parents in her bid to become ruler), after which he travels through a time zone into the 20th Century, becomes a member of the French Secret Service (F.R.O.7.) as Freddie the Frog, and is then sent by them to Britain in order to foil a plot by arch-villain El Supremo and Messina to enslave the world's population. Somewhere in amongst all of this mayhem, Freddie encounters Nessie (voiced by Phyllis Logan) and her many other long-neck relatives in Loch Ness, and she informs him in song (and dance) of what a shy girl she really is. An excerpt from the film that includes this song, and which in my opinion is both the most entertaining and the most beautifully animated section of the entire film, can be viewed here: https://www.youtube.com/watch?v=hjQns22P99g

Pater Moeskroen (1992). 'Nessie'.
This is a Dutch folk band but their music is also infused with Celtic, klezmer, and punk elements. Their LNM song, from an early 1990s album, is apparently all about Nessie's, ahem, intimate liaisons with other loch monsters – but I don't speak Dutch, so I wouldn't know...
https://www.youtube.com/watch?v=XFOedIU03iM

Some Velvet Sidewalk (1992). 'Loch Ness'.
American experimental lo-fi rock band Some Velvet Sidewalk released two different versions of their Nessie-themed song 'Loch Ness'. One version appeared as track #2 on their own 1992 album, 'Avalanche'; the other had appeared a year earlier, again as track #2, but this time on 'Kill Rock Stars', which was a compilation album featuring a number of different acts.
https://www.youtube.com/watch?v=eBZUrh_dwsM

Those Darn Accordions (1996). 'Deathbed Confession'.
This song from San Francisco accordion band Those Darn Accordions' 1996 album 'No Strings Attached' takes its inspiration from the 'deathbed confession' claim of Christian Spurling in 1993 regarding his supposed (but never confirmed) hoaxing of the famous Surgeon's Photograph, by having allegedly made a head-and-neck model of Nessie, attached it to a toy submarine, and set it afloat on Loch Ness one day in April 1934, where it was then reputedly photographed, yielding the iconic image that gynaecologist and purported co-conspirator Robert Kenneth Wilson then passed off to the media as a genuine Nessie photo.
https://www.youtube.com/watch?v=7TlxXWDCz-4

Jon Downes and The Amphibians From Outer Space (1995). 'Invocation of My Daemon Brother'; (2000). 'The Loch Ness Monster'.
It is only fitting that a cryptozoologist should have contributed some Nessie-featuring songs, and in the case of Jonathan Downes, founder of the Centre for Fortean Zoology (CFZ) but also an accomplished musician with his backing band The Amphibians From Outer Space, he has two to his name. The legendary Tony 'Doc' Shiels appeared on the opening track, 'Invocation of My Daemon Brother', from Jon's 1995 album 'The Case', proclaiming: "I don't believe in the Loch Ness Monster, I don't believe in the Loch Ness Monster, even though I seen the [expletive] and photographed it! But how the hell can anybody in their right mind believe in the Loch Ness monster, what do you mean, right mind?". Jon's second entry in the Loch Ness jukebox came in 2000 from his album for Voiceprint records, 'The Weird World of Lionel

Fanthorpe and Jon Downes and the Amphibians From Outer Space', on which another Fortean legend, Father Lionel Fanthorpe, sang 'The Loch Ness Monster', a ditty about Nessie featuring a punked-up version of the old Irish song 'The Wearing of the Green'.
http://www.cfz.org.uk/music/case/01-AudioTrack%2001.mp3

The Real McKenzies (2001). 'Nessie'.
This song is the opening track to Canadian Celtic punk band The Real McKenzies' 2001 album 'Loch'd and Loaded', and is basically a protest song regarding the search for and potential future capture of Nessie.
https://www.youtube.com/watch?v=Ps6Stqlg70A

Judas Priest (2005). 'Lochness'.
This lengthy track (13.28 minutes long) is the tenth and final one on world-famous British heavy metal band Judas Priest's fifteenth studio album, 'Angel of Retribution', and was co-written by their legendary leather-clad frontman, Rob Halford, returning to the band after an absence of 15 years.
https://www.youtube.com/watch?v=9ncnKjZzOPs

Honorary mentions are also due to the following three songs, which are not themselves about Nessie but feature Nessie-themed official videos in the first two instances and in the third instance is the title track of an entire Nessie-themed movie:

Reggie and the Full Effect (2005). 'Get Well Soon'.
The video to this song, track #2 from Kansas City rock band Reggie and the Full Effect's 2005 album 'Songs Not To Get Married To', features the collapse and total disintegration of a green, suspiciously arm-puppet-like Loch Ness monster's entire life, beginning with a savage divorce settlement in which he loses his loch and is forced to roam the streets homeless as his life falls apart, reduced to living in cardboard boxes. Unrelentingly dark and grim, there is no happy ending for this video's LNM.
https://www.youtube.com/watch?v=kGW1ceFeNfE

Judas Priest lead singer Rob Halford – not only a fellow biker and a fellow West Midlander (we were born just a few miles from one another) but also, it would seem, a fellow Nessie enthusiast! (© Rob Halford/Kerrang!)

176

The Automatic (2006). 'Monster'.
In pleasant contrast, the video to this 2006 song, track #5 on Welsh rock band The Automatic's debut album 'Not Accepted Anywhere', is pure slapstick comedy, featuring a Nessie whose vertical neck rising above the loch's surface is of veritable skyscraper proportions yet somehow still manages to go unnoticed by the band, starring here as hapless monster seekers. So too does a dancing bigfoot that definitely gets down and gets with it as their song plays, before things finally hot up in every sense for our heroes when they have an exceedingly close encounter with a UFO.
https://www.youtube.com/watch?v=5x29ug8gV5M

Adam Faith (1961). 'What a Whopper'.
Also well worth including here is this title track to a British b/w comedy film from 1961 entitled *What a Whopper*, featuring an attempt by a young struggling would-be novelist to raise money by writing a book about the Loch Ness Monster and then, to generate plenty of publicity for it and thus ensure its success, staging a hoaxed Nessie sighting - only for the real Nessie herself to make a surprise, and very tongue-in-cheek, appearance in the closing scene of the film. It starred British rock 'n' roll singer and actor Adam Faith, who also sang the toe-tapping theme song (written by Johnny Worth) that opens the film:
https://www.youtube.com/watch?v=CgWlOpsuuPw

Loch Ness (public domain)

177

In addition to those songs documented here, in which Nessie features extensively in the lyrics, there are a fair few others in which she is mentioned briefly or in passing. No less than 57 of these, recorded by the likes of Eminem, Roger Taylor, Pras, De La Soul, and Crash Test Dummies among others, can be accessed at this site: http://www.lyrics.net/lyrics/loch%20ness%20monster

Finally: Although by far the most extensively represented example, Nessie is not the only water monster to have inspired various songs and other musical compositions. Here are three notable non-LNM examples:

THE LAKE OKANAGAN MONSTER: Paul Whiteman Orchestra (1924). 'The Ogo-Pogo – The Funny Fox-Trot'.
As every self-respecting cryptozoological enthusiast will readily confirm, this is the English music-hall song from 1924, composed by Mark Strong, that subsequently gave its name to the now-famous water monster of Canada's Lake Okanagan (until then, it had been known only as the naitaka - a traditional Native American name given to it by the local Okanakane nation). Despite the song featuring a banjo-playing terrestrial monster from Hindustan (additionally sporting a pair of antennae and wearing boots in the delightful illustration by Fred Low adorning its sheet music's front cover) - far removed indeed from Canada's unequivocally non-musical aquatic cryptid of serpentiform shape – the name Ogopogo stuck, and the Lake Okanagan monster has been affectionately referred to by it ever since.

Further details concerning this song – including how I was fortunate enough to encounter and purchase a copy of the sheet music for the Savoy Havana Band's original version of it from 1924, thereby enabling me to include its front cover illustration for the very first time in a cryptozoological publication (my book *In Search of Prehistoric Survivors*, 1995) -

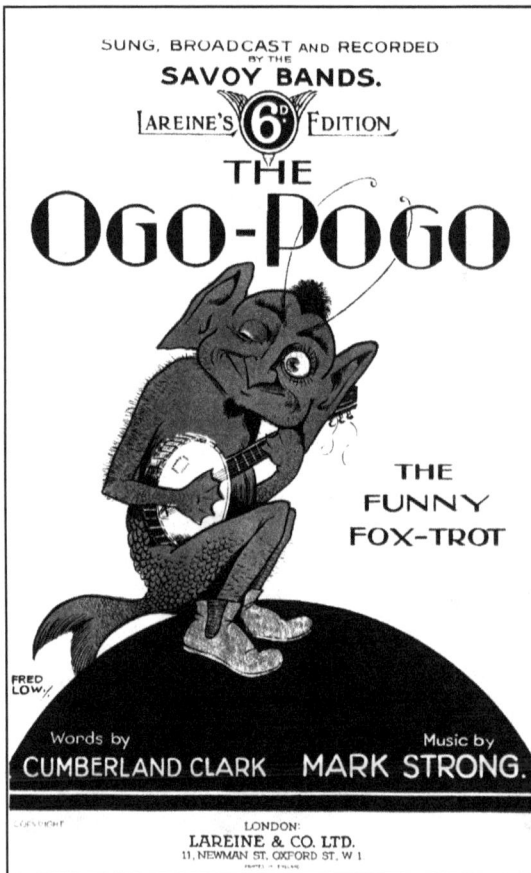

Front cover of my original copy of the sheet music for the Savoy Havana Band's version of 'The Ogo-Pogo – The Funny Fox-Trot' (© Dr Karl Shuker)

can be found in my book *The Menagerie of Marvels* (2014). Several different acts released this song in 78 rpm record format during the 1920s, including the Paul Whiteman Orchestra in 1925 that featured Billy Murray as vocalist singing reworded American lyrics (which can be listened to via the link below) rather than the original English ones written by Cumberland Clark, the Savoy Havana Band singing the original English lyrics (see illustration facing), Meyer Davis' Swanee Syncopaters, and George Berry (aka Harry Fay).
https://www.youtube.com/watch?v=uQE8T6Ip6Ic

THE PATAGONIAN PLESIOSAUR: Arturo Terri (1922). 'El Plesiosaurio Tango'.
Seemingly as elusive as the lake-dwelling Patagonian long-neck that it celebrates, this exotic-sounding crypto-composition has evaded every attempt not only by me but also by several friends and correspondents on Facebook to track down an online recording of it – but this is only fitting, I suppose, bearing in mind that its subject also succeeded in remaining concealed from those searching for it. With lyrics by Amilcar Morbidelli, it was composed in 1922 by Rafael D'Agostino to commemorate Argentinian biologist Dr Clemente Onelli's expedition during April of that same year to a mountain lake near Esquel in Argentina, seeking the so-called Patagonian plesiosaur that had allegedly been sighted there by an itinerant Texan adventurer called Martin Sheffield who had lived off the land in Patagonia for many years. Sponsored by Buenos Aires Zoo, of which Onelli was the director, the expedition did reach this lake, but no sightings of cryptids were made (further details can be found in my book *In Search of Prehistoric Survivors*, 1995). D'Agostino dedicated his plesiosaur tango to Onelli, a caricature of whom is humorously portrayed riding the plesiosaur on the front cover of the Arturo Terri version of this composition's sheet music (see this present book's Nessie picture gallery).

If anyone reading this chapter knows of an online recording of 'El Plesiosaurio', I'd greatly welcome details. Meanwhile, here is a link to its Spanish lyrics:
http://www.todotango.com/musica/tema/6341/El-plesiosauro/

Incidentally, this was not the only 1920s musical composition inspired by Onelli's Patagonian pursuit of plesiosaurs. Here is the delightful front cover illustration from Fernando Randle's piano sheet music for his own composition, 'El Plesiosauro Tango' (note the slight difference in its title's spelling from that of D'Agostino's tango), featuring a very dapper pipe-smoking plesiosaur with top hat, spats, and cane! Sadly,

Front cover of the piano sheet music for Fernando Randle's 'El Plesiosauro Tango' (public domain)

179

however, Randle's plesiosaur tango was not as popular as D'Agostino's. Once again, I haven't been able to locate an online recording of it, so I'd greatly appreciate any assistance in doing so.

Julio Fava Pollero's 'Antediluvian Tango' was a third tango inspired by Onelli's plesiosaur hunt, but although he performed with his own orchestra he never released this composition in record form, only as sheet music, published in 1927. This was because by then the swell of public interest in the Patagonian plesiosaur expedition had subsided. Its sheet music's front cover depicted a humorous caricature of Onelli attempting to tie the plesiosaur down.

My sincere thanks to several Facebook friends, especially Karl J. Claridge, Claudio Diaz, Adam Naworal, Jeff Rausch, and Valerie Wyllie, for supplying me with information and images relating to this trio of Patagonian plesiosaur tangos.

THE GREAT SEA SERPENT: Maurice Strakosch (1850). 'Sea Serpent Polka'.
Inspired by a bout of sea serpent sightings off the towns of Gloucester and Nahant in Massachusetts, New England, USA, during 1817-1819, this very jaunty polka was written in 1850 by Maurice Strakosch, an American musician of Czech origin, and featured an immense snake-like sea serpent coiled upon the front cover of its sheet music. An undated recording of concertina player Michel Van Der Meiren performing this lively composition can be listened to on YouTube:
https://www.youtube.com/watch?v=37rLMOpuJv8

Many thanks to Facebook friend Jeff Meuse for bringing this charming instrumental composition to my attention.

I do hope that you've enjoyed this very special Nessie concert - if so, then that is definitely music to my ears!

Front cover of the sheet music for 'Sea Serpent Polka' (public domain)

CHAPTER II:
'THE LOCH NESS MONSTER'
- A POEM

This wouldn't be a book of mine if cryptozoology didn't rear its head in it somewhere! So here, for mystery beast lovers everywhere, is a little snippet of crypto-history – the very first piece of writing that I ever penned (over 40 years ago!) in relation to the most famous cryptid of all. Nessie, this one's for you!

Karl Shuker – *Star Steeds and Other Dreams: The Collected Poems*

Writing poetry has been a longstanding hobby of mine, and here is that above-mentioned snippet - my tribute in verse to Nessie:

THE LOCH NESS MONSTER

Through swirling mists of early morn,
Across the loch's dull light,
A dark shape moves with hidden power,
Then disappears from sight
Beneath Loch Ness's murky cloak
Of water black and cold.
What lives in this vast underworld,
In Ness's misty fold?

What creature thrives below the waves,
Beneath the surface grim?
What beast appears in photographs –
Obscure, opaque, and dim?
What creature gave the myths and tales
From bygone days new fame?

Of water horses, fierce, malign,
Which from the waters came,
To strike the hearts of every man
With terror of their forms.
The kelpies – dark, malignant ghosts,
And harbingers of storms.

So what gave all these legends life –
A creature huge in size,
With tiny horns and rhomboid limbs,
And glowing, deathless eyes?
A relic from the ancient past,
Disturbed by modern worlds?
It will be long, indeed, before
This secret is unfurled.

"Beneath Loch Ness's murky cloak of water black and cold. What lives in this vast underworld, in Ness's misty fold?" (illustration © Richard Svensson)

TAKING A SHINE TO NESSIE RECEIVES THE STAMP OF APPROVAL – THE LNM ON AN FDC

Writing poetry, incidentally, is just one of several longstanding hobbies of mine. Another one is philately, which my case takes a thematic approach, because I collect postage stamps specifically featuring animals – including mystery animals.

Yet despite having been named in various polls as Scotland's top tourist attraction, Nessie has never been commemorated on a British postage stamp – or has she? Well, sort of...

On 20 August 1991, the Royal Mail issued a set of five commemorative postage stamps depicting the skeletons of various dinosaurs. It also produced in association with this set a

Proudly displaying my copy of the Adrian Shine-signed Nessie FDC
(© Dr Karl Shuker)

First Day Cover (FDC), released in Inverness on that same date. This FDC features not only all five dinosaur stamps but also a beautiful full-colour illustration of the LNM, depicted in its most popular, plesiosaurian reconstruction and headed 'The Monster of Loch Ness'.

I'm not at all sure what palaeontologists must have thought about cryptozoology's most controversial megastar sharing equal billing with the likes of such dinosaurian celebrities as *Iguanodon*, *Stegosaurus*, *Tyrannosaurus*, *Protoceratops*, and *Triceratops*. However, worse – at least for them – was still to come.

The FDC also bore two ornate circular franks, both of which took the form of Nessie as a plesiosaur encircled by the words "The Last Dinosaur? The Loch Ness Monster". Now, as any schoolboy dinosaur enthusiast would be quick to point out, plesiosaurs were definitely not dinosaurs, belonging instead to an entirely separate superorder of prehistoric reptiles, Sauropterygia. So even if Nessie does exist, and is a living plesiosaur, she is certainly not the last – or indeed any other – dinosaur.

Never mind. This taxonomic tribulation was more than compensated for in my eyes when, after being fortunate enough to stumble upon and duly purchase one of these highly attractive FDCs at a collector's fair several years ago, I discovered to my delight that it had been personally signed by none other than veteran Nessie investigator Adrian Shine, alongside a blue-lettered frank that read: "Carried aboard the research ship Ecos on Loch Ness in the search for Nessie and signed by the project leader Adrian Shine". Check it out here in colour: http://karlshuker.blogspot.co.uk/2013/08/taking-shine-to-nessie-receives-stamp.html

All in all, a perfect addition to the collection of any cryptozoological philatelist!

And while on the subject of LNM-inspired philately, Nessie was also featured, again in plesiosaurian guise, on a spectacular miniature sheet issued by the Maldives in 1992.

For details of many additional cryptozoologically-themed postage stamps, check out my book *Dinosaurs and Other Prehistoric Animals on Stamps: A Worldwide Catalogue* (2008), which includes a special appendix listing every cryptozoology stamp that had been issued up until that time.

Maldives miniature sheet depicting Nessie, issued in 1992, from my personal collection (© Maldives Royal Mail Postal Service)

CHAPTER 12:
KELPIES AND ⊕THER WATER-H⊕RSES

Faster and faster went the pony, and, as it went, the wind began to rise again and the clouds to come chasing across the sky. Instead of going towards the pond, the pony made for the sea-shore, and they [the children riding it] could hear the waves moaning on the beach.

"How I wish I'd gone straight to see my blue-eyed sweetheart!" exclaimed the eldest brother. "Let's stop the pony, and we'll race back before the storm breaks."

But he tugged and pulled in vain; the pony galloped faster and faster, neighing with glee when it came in sight of the waves.

Over the beach it went and straight into the foaming billows, with the children clinging in terror to each other and to its mane...

Further and further into the cruel, raging sea went the pony; and at last the waves closed over them all.

Roger Lancelyn Green – 'The Goblin Pony',
in *Myths From Many Lands*

Long before modern-day cryptozoologists speculated that Ness and various other Scottish lochs may harbour elusive plesiosaurs, long-necked seals, giant eels, and/or sundry other exotic fauna of the corporeal kind, traditional Highland folklore claimed that some of these brooding bodies of freshwater were home to fearsome supernatural entities known as water-horses.

Various kinds of water-horse have been delineated and named, based primarily upon their geographical location and the type of freshwater abode that they reputedly frequent. Of these kinds, the most formidable and dreaded is undoubtedly the each uisge (pronounced 'eck ooshkya'), which haunts the Highlands' lochs and the sea. In contrast, the Scottish kelpie is linked to rivers, streams, fords, waterfalls, and other sources of running water. The Isle of Man has its very own, unique type of water-horse, called the cabyll-ushtey.

Nor are these malevolent beings limited to Scotland and the Isle of Man within the British Isles. Ireland has its own equivalent entity, which is known as the pooka; there is also the Irish

Keep well away from kelpies! (public domain)

aughisky. And Wales's counterpart is the ceffyl dwr, notorious for leaping on the backs of weary travellers.

As their name indicates, the most common guise assumed by water-horses and also the pooka is that of a horse or pony, usually black in colour (but pale grey in the case of the Manx cabyll-ushtey), with rough, shaggy, unkempt hair and mane, usually wet or damp to the touch, plus a faintly stagnant odour, and glowing, demonic eyes. In addition, if observed closely its hooves will be seen to be reversed. Such a steed will attempt to entice unwary humans, especially children, to mount it and be taken for a ride. But if they are reckless enough to do so, they find themselves unable to dismount, having instantly become stuck fast to its back (which magically lengthens to accommodate any number of persons riding it). They can then do nothing other than watch in abject, impotent horror as the predatory water-horse immediately races directly into its aquatic domain and plunges down into the depths, promptly drowning and then greedily devouring its hapless, helpless riders.

Sometimes, a water-horse will be ensnared by a canny farmer using a halter stamped with the Sign of the Cross, and is then harnessed to a plough alongside a team of mortal horses. However, its supernatural strength is such that it will readily haul plough and horses alike along with it as it races off into its welcoming loch or river, where it soon shakes off the plough and tears apart the doomed horses.

Having said that, there are folk stories of water-horses mating with normal horses. Their resulting hybrid progeny can never drown, and can be physically distinguished from pure-bred normal horses by their extremely short ears.

Very occasionally, a water-horse is actually killed, by being shot with a silver bullet or by being stabbed with fire-heated spears forged out of iron, but no corpse or carcase is ever left behind. Nothing remains at all, in fact, other than a pool of water, or a jelly-like substance very reminiscent of so-called star rot or pwdre ser (itself a highly mysterious gelatinous material of undetermined nature sometimes found on grass and foliage, especially after rain, and for which several different identities have been suggested, but never confirmed, including slime moulds, cyanobacteria, and frog spawn devoured but then vomited forth in semi-digested state by predators).

If chased by a kelpie, one certain means of eluding it is to jump over a stretch of river or stream. Even if it is only very narrow, this will still be sufficient to hold back the kelpie, because, strangely enough, it is unable to cross any stretch of running water, despite the fact that it can emerge from or enter into any such stretch.

Although it occurs more commonly in its equine guise, the shape-shifting water-horse will sometimes assume human form. It usually appears as a tall, thin youth or young man, whose clothes seem damp, as does his long black hair – which if observed very closely can be seen to contain strands of water weed and grains of sand. He usually wears boots, to conceal the fact that even when he is adopting a human guise, his feet remain hoofed. In the case of the pooka, conversely, its human form is usually a wizened, toothless old man, with evil leer and flashing eyes.

As a child, I was lucky enough to receive as gifts from my family a series of large-format hardback books of world myths, legends, and folktales vividly retold by eminent folklorist Roger Lancelyn Green, beautifully illustrated by Janet and Anne Grahame Johnstone, and published by Purnell. One of these volumes, *Myths From Many Lands* (1965), included Green's retelling of a traditional Breton folktale, which he entitled 'The Goblin Pony' (a short excerpt from which is quoted at the beginning of this present chapter). Years later, I discovered that in both appearance and behaviour, this folktale's eponymous supernatural entity was identical to the Irish pooka, and so fascinated me that a few years ago I penned my own, greatly-expanded version of it, which I reset in Ireland. Now, however, I am relocating it again, this time to Scotland, and replacing the pooka with one of its equally dangerous Caledonian counterparts, the kelpie

As noted above, the kelpie's most innocuous yet deadliest guise is a dark shaggy-coated colt or pony of deceptively playful, harmless demeanour. On first sight, it is easy to mistake a kelpie for a genuine animal - until you see its eyes, which betray its true identity by blazing with a scorching, unholy fire. Consequently, it is always best to avoid anything that might be a kelpie - otherwise, as I now reveal in my retelling of an age-old Scottish folktale, you may not live to regret your mistake:

COLT OR KELPIE?

It was Halloween, so although she was a weaver of cloth by trade, tonight Molly was spinning tales of wonder and magic instead, regaling her three grandchildren with stories of the kelpie, the boobrie, the green-furred cu sith or fairy hound, and many other mythical Scottish entities - but their attention was beginning to waver. Kevin, aged 15, was the eldest of the trio, and had

Beware the deceptively pony-like but decidedly malevolent water-horse! (© Andy Paciorek, from *Strange Lands* - http://www.blurb.co.uk/b/1957828-strange-lands and http://www.batcow.co.uk/strangelands/)

thoughts only for Kitty - the fair-haired daughter of their village's new school teacher. Kevin's sister, Jeannie, just a year younger, was mentally designing the new dress that she planned to wear at the forthcoming autumn fete. Only Alex, their six-year-old brother, was still listening to Molly's words, although the log fire was gently lulling him to sleep.

Molly smiled, pausing in her story-telling; and, as she did so, Kevin stood up. The night outside was warm, and he had decided to take a walk, secretly hoping that Kitty might be doing so too, and that they may then happen to meet. His plans were swiftly dashed, however, because Jeannie and Alex wanted to accompany him - so instead of walking into the village, they wandered through the outlying meadows and forests.

Encouraged by the evening's pleasant ambience, the three youngsters strolled further than originally intended, eventually approaching the banks of a very deep, dark, fast-flowing stretch of river that they had never visited before - and with good reason. Traditionally, the villagers kept away from this area, because they believed it to be an accursed place - haunted by sinister shadowy forms.

The river raced onwards before them, an ebony stream of liquid obsidian beneath a cloudy sky - and when the three youngsters stared into its sable depths, an additional flurry of ripples abruptly raced across its surface, as a cold wind began to harry the clouds above. Jeannie shuddered, and they were about to turn back for home when, to their great surprise, a frisky pony-like colt suddenly appeared, as if from nowhere, and began frolicking in front of them, at the river's edge.

It was an uncommonly hairy colt, and as black as the waters of the river. For a few moments, it refused to let the children draw near, playfully gambolling out of reach as soon as they stretched out their hands to pat it, but it soon quietened, allowing them to stand beside it and stroke its rough, unkempt coat.

Where had this colt come from? It had neither bridle nor saddle, and even its hooves were unshod. Whatever its origin, however, it must certainly have been splashing in the river, because its coat was very wet - but the evening was warm, so who could be surprised?

Nevertheless, for some indefinable reason Kevin felt strangely ill at ease concerning their new-found friend. It was obviously just a trick of the light, but sometimes the colt's long mane appeared to him to be almost alive, dancing even when the breeze had stilled. And occasionally, when its eyes met his own unexpectedly, just for an instant it seemed to Kevin as if they were infused with fire - as if bolts of lightning flickered in their fathomless depths.

I've been listening to too many of Granny Molly's fairy stories, he thought to himself, smiling wryly as the colt rubbed itself against his legs, whinnying with pleasure as his fingers rubbed its ears. Suddenly, hardly knowing what he was doing, Kevin leapt onto its back, his strong legs gripping its damp flanks, and his hands securing a firm hold around its powerful neck.

Without further ado, Jeannie mounted behind him, and there was just enough room for Alex to sit behind her. When all three were seated, Kevin gently tapped the colt's flanks with his heels, and it began to walk slowly across the meadow.

Laughing, they urged it to go faster, and, obligingly, the colt quickened its pace, from a canter into a trot - and then from a trot into a gallop. Alarmed, Kevin tried to slow the animal down, pulling at its mane, and calling out, but to no avail. On and on it ran, too fast for them to make any attempt to jump from its back. All that they could do was hold on and hope that it would eventually tire. Instead of tiring, however, it seemed to grow ever stronger.

Suddenly, the colt changed direction - to their horror, it was now heading directly for the river! Frantically, they tried to stop it, but it was too late. Even as they opened their mouths to scream, the creature had reached the water's edge - and as it plunged into its depths, the river seemed to rise up on all sides, welcoming the return of its demonic denizen, and engulfing the hapless victims that it had abducted.

The hours had seemed like a lifetime to Molly, as she searched the village and surrounding countryside for her missing grandchildren. No-one had seen them, and no-one knew where they may be. Now, much to her apprehension, she was nearing the ill-omened stretch of river that everybody had shunned for as long as anyone could remember. Old legends die hard, thought Molly, and some are more than just legends - especially on the night of Halloween.

At that same moment, a shadow moved just ahead, but as she turned to look at it Molly realised that it was not a shadow. It was a tall, thin man, whose clothes looked wet, and whose long, black, unkempt hair seemed to be matted with water weeds. He gazed briefly at Molly, an evil grin stretching almost from ear to ear, but far worse were his eyes, for as they stared into hers they seemed to glow red, like cauldrons of scarlet flame. Molly shied away in terror, but when she looked back again the apparition was gone, as if it had never been.

Only the cold silver moonlight remained, lighting up the desolate landscape ahead, where now, at the river's edge, Molly could see three long boulders - or so she thought, until she walked closer. Her heart seemed to explode as she looked down at them - the 'boulders' were her grandchildren, lying cold and grey and still. They had drowned - all three of her grandchildren were dead.

Numb with shock, for a time she was unable to tear her gaze away from this terrible scene, but when she finally looked up she realised that something was moving close by. Out of the night's shadowy darkness, something even darker was emerging. It looked like a pony, a black shaggy-coated colt, throwing its head upwards and neighing as it ran towards her.

Such a merry creature was an incongruous contrast to the tragedy that she had just witnessed, but then, with flaring nostrils and dancing mane, the colt stared up at her - and Molly gasped in horror. Its exultant eyes were suffused with evil, and blazed like glowing coals, aflame with the very fires of Hell! This was no harmless colt - it was a kelpie!

Molly staggered back, almost stumbling to the ground, and the kelpie reared up triumphantly, its forelegs lashing out at her with razor-sharp hooves. As she fell, however, she dropped her bag, whose contents tumbled out onto the grass. Among them was her Bible, which she instantly grasped, thrusting it up towards the malign beast's face. As she did so, the moon's rays illuminated the bright golden Cross emblazoned upon its cover - and the kelpie disappeared.

CHAPTER 13:

LESSER NESSIES
‒ SURVEYING THE 'OTHER' MONSTERS OF MAINLAND SCOTLAND'S FRESHWATER LOCHS

> Morag, Harbinger of Death,
> Giant swimmer in deep-green Morar,
> The loch that has no bottom...
> There it is that Morag the monster lives.
>
> Anonymous – Traditional Scottish song

Everyone has heard of Nessie, the reputed monster of Loch Ness, but fewer people realise that mystery beasts of various forms have also been reported from a sizeable number of other mainland Scottish freshwater lochs. Many of these reports were first compiled in Peter Costello's standard work *In Search of Lake Monsters* (1974) and later summarised in Michael Newton's very comprehensive *Encyclopedia of Cryptozoology* (2005), but here is a representative selection.

LOCHS ARKAIG, ASSYNT, AND FEITH AN LEÒTHAID

With a maximum depth of 359 ft and measuring 12 miles long, Loch Arkaig is situated in the Lochaber area of the Highlands. In a diary entry for 3 October 1857, English politician Lord Malmesbury recorded that his game stalker, John Stuart, had twice seen at Achnaharry the horse-like head and hindquarters of a 'lake-horse' basking at the loch's surface at sunrise when there were no ripples on the water. This loch monster has since been dubbed Archie.

In an article on Scottish loch monsters published during the early 1850s, Malmesbury included a claimed monster sighting from 1837 on Loch Assynt in Sutherland by two fishermen, who also saw it a second time shortly afterwards on a small island in this 6.3-mile-long loch. Very hairy, and grey in colour, the creature was compared by them to a young bull in size but with a broader back. It was about 3 ft tall, quadrupedal, with a bulldog-like head and large eyes.

Loch Feith an Leòthaid is connected to Loch Assynt, and during the 1930s an unidentified creature with a long neck and a deer-like head apparently surfaced close to the boat of Kenneth MacKenzie from Steen, gazing across this vessel's stern before disappearing again.

LOCH AWE
The third largest freshwater loch in Scotland by surface area (which is approximately 15 square miles), Loch Awe in Argyll and Bute is also this country's longest at 25 miles in total, and is reputedly home to a mysterious serpentiform monster known as the beathach mór. As far back as the 16th Century, fishermen were claiming that this loch's waters harboured gigantic eels "as big as a horse with an incredible length" - a belief that remains prevalent here today, though no eel of such inordinate dimensions has ever been drawn forth and made available for scientific scrutiny.

Vintage picture postcard depicting Loch Awe (public domain)

LOCH EIL
One of the most unusual water monsters reported from a Scottish freshwater loch is the faceless, vermiform horror encountered at Loch Eil in the western Highlands by author Denys

James Watkins-Pitchford (better known by his nom-de-plume "BB") and documented by him in 1962. Here, quoted directly from his book *September Road to Caithness and the Western Sea*, is his first-hand description of what he saw:

> I was watching some mallard paddling about among some weedy rocks at the end of a little promontory when there appeared out of the calm water exactly opposite me a large black shiny object which I can only compare with the blunt, blind head of an enormous worm.
>
> It was, I suppose, some 50 yards from where I was standing, and it kept appearing and disappearing, not moving along, but rolling on the surface. The water was greatly disturbed all round the object. It had a shiny wet-looking skin, but the head (if head it was) was quite unlike a seal's and had no face, or nose, no eyes. It rose quite a long way out of the water, some three feet or more, before sinking back.

The most obvious explanation for a large elongate creature in a Scottish freshwater loch is an eel; but unless the creature was inaccurately recorded by its eyewitness, an eel with no face, not even any eyes, would be a very unusual one indeed - and one that could rise 3 ft or more out of the water would be even more so (as would a worm for that matter!).

Incidentally, Loch Eil is linked to Loch Linhe, a sea loch on Scotland's west coast and where, during the 1890s, a still-unidentified eel-like animal of sizeable length but bearing a mane was found dead at Corpach Lock, close to Fort William at Linhe's north end. Might this have been

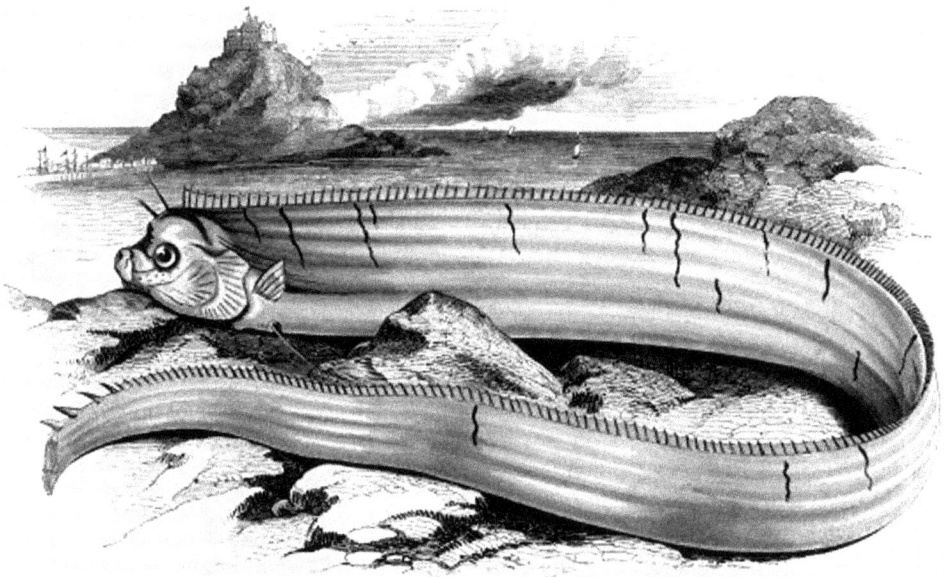

Vintage engraving of a beached giant oarfish (public domain)

a vagrant giant oarfish *Regalecus glesne* that had made its way, or (perhaps dying) had been carried by water currents, into this coastal loch from the open sea? Long-necked Nessie-type monsters have also been sighted here, during the 1940s and again during the 1960s.

LOCH GARTEN

Situated in the Strathspey area of Scotland's Cairngorms National Park, Loch Garten is most famous nowadays for the RSPB-coordinated success story in the breeding of wild ospreys here, but in bygone times it was famed for reputedly being home to a fearsome lake monster known as a water-bull. Resembling a hybrid of horse and bull, it sported a huge head, a jet-black mane, and would give vent to an extremely loud, hideous roar.

According to local lore, a bold crofter once sought to trap this formidable creature, using as bait a young lamb attached to a very large hook, which in turn was tethered by a long sturdy rope to a huge lochside boulder weighing many tons. After rowing out to the centre of the loch and dropping the hooked lamb there, the crofter returned to shore in the hope that the water-bull would swallow the bait during the night, and thus be snared internally by the engulfed hook, after which he would haul the beast ashore. But when he checked the following morning, both the lamb and the boulder were gone. All that could be seen was a deep rut in the ground, where something with immense strength had dragged the massively heavy boulder into the loch.

As the water-bull was never seen or heard again, the inference in this tale is that once in the water, the huge boulder's weight had dragged the water-bull down to the loch bottom - where, unable to free itself from the hook that had snared it internally when it swallowed the lamb, the monster had drowned.

LOCH LOCHY

The lesser Nessie that has attracted most media attention in fairly recent times is Lizzie, the monster of 10-mile-long Loch Lochy - Scotland's third deepest loch (531 ft at its maximum depth), sited immediately below Loch Oich. With no publicised sightings for 36 years, Lizzie reclaimed the headlines in September 1996, when a 12-ft-long, dark-coloured mystery beast with a curved head and three humps reared up out of the water and began moving round in circles in full view of several staff and guests at the Corriegour Lodge Hotel, overlooking the loch. According to Aberdeen University psychology student Catriona Allen, who studied this amazing sight through binoculars: "It certainly wasn't a seal, otter, porpoise or dolphin".

In late July 1997, a six-man expedition featuring previous Loch Morar diver Cameron Turner and led by Gary Campbell, president of the Official Loch Ness Monster Fan Club, arrived to conduct a sonar sweep of the loch. Encouragingly, they achieved success on their very first day, when their equipment detected a large unidentified object swimming in the middle of the loch and estimated at 15-20 ft long - far bigger than anything known to be there. Turner came back to Lochy in September 1997, but no new evidence was obtained.

LOCH LOMOND

By surface area, totalling 27 square miles, Loch Lomond in Scotland's West Dunbartonshire/ Argyll and Bute/Stirling region, and marking the boundary between central Scotland's highlands and lowlands, is the largest stretch of inland water in the whole of the island of Great Britain. In terms of anomalous aquatic animals and other esoterica, moreover, it is also famous as the locality claimed in an atlas published in 1659 to harbour "fish without fins" and a mysterious "floating island". And in 1724, Alexander Graham of Duchray claimed that locals living nearby sometimes see the water-horse reputedly inhabiting its waters.

More recently, at Easter 1980, a Mr and Mrs Maltman and their daughter were camping near the edge of Loch Lomond at Luss when a head and slender neck rose up to a height of about 5 ft above the water surface, no more than 200 yards away, with a long curved back visible behind. This amazing spectacle lasted for 30 seconds or so, then the head and neck swiftly submerged and were not seen again. The Maltmans were so frightened that they fled, later returning only to pack their belongings before journeying back home. And in 1997, a somewhat indistinct, unidentifiable moving object was filmed in the loch by investigator Nick Taylor.

Vintage postcard depicting Loch Lomond and its islands (public domain)

Equally unexpected but totally verified, incidentally, is the presence on Inchconnachan, one of this loch's islands, of a naturalised, thriving population of Australian red-necked wallabies

Macropus rufogriseus. They are descended from some that were introduced there during the 1940s by Lady Arran Colquhoun, and Inchconnachan is nowadays referred to colloquially as Wallaby Island.

LOCH MAREE (aka LOCH NA BÈISTE)
The fourth largest of Scotland's freshwater lochs by surface area, and situated in Wester Ross in the Western Highlands, Loch Maree is also referred to as Loch na Bèiste ('Loch of the Beast' in Scottish Gaelic), due to the muc-sheilch. This is a local name popularly applied to its own particular water monster and loosely translates as 'turtle-pig'. Yet despite its descriptive name, and the fact that sightings of this monster are reminiscent of Nessie reports, featuring humped backs rising above the surface and resembling capsized boats, zoologists have sought to identify it as merely a large eel.

LOCH MORAR
The most famous lesser Nessie is Morag, the monster of Loch Morar, whose history, like Nessie's, dates back many centuries, as testified by the very old Scottish song quoted in part at the beginning of this present chapter.

Loch Morar is 11 miles long, approximately 1.5 miles wide, and with a maximum depth

Vintage postcard depicting Loch Morar (public domain)

exceeding 1000 ft it is Britain's deepest freshwater lake. Unlike the waters of Loch Ness, however, which are extremely peaty, Morar's are very clear, enabling objects situated at quite a distance beneath the surface to be perceived with remarkable clarity - as exemplified by visitor Robert Duff's extraordinary sighting on 8 July 1969.

A joiner from Edinburgh, Duff was fishing from a boat in Meoble Bay on the loch's southern shore, where the water is no more than 16 ft deep and very lucid, when he spotted what he described as a "monster lizard", lying motionless on the loch's white, leaf-strewn bottom, looking up at him. Duff estimated the creature to be 20 ft long, with a snake-like earless head, slit eyes, and a wide mouth. Its body was grey-brown with rough skin, and it had four limbs, with three toes visible on each front foot, plus a tail. He was so startled that he revved the boat up and made off at once. Later, however, he returned to the same spot, but the animal had gone.

Even more dramatic was the 5-minute confrontation experienced on 16 August 1969 by Duncan McDonell and William Simpson. At about 9.00-9.30 pm, but while still daylight, their motor boat was travelling along the loch at a speed of 6-7 knots when McDonell, at the wheel, saw a creature in the water about 20 yards behind but moving directly towards them. A few seconds later it caught up, and collided with the side of their boat, seemingly unintentionally but nonetheless with sufficient force to hurl a kettle of water off the boat's gas stove and onto the floor. McDonell attempted to fend the beast away with an oar, frightened that it may capsize the boat, but because the oar was old it snapped in half.

When Simpson saw this, he picked up his rifle, ran out of the cabin, and aimed a shot at the creature - which slowly sank away from the boat. They did not see it again, but they did not see any blood either, or any other sign to indicate that Simpson's bullet had hit it.

According to Simpson and McDonell, the portion of the creature that they had observed was 25-30 ft long, with rough, dirty-brown skin, and three humps or undulations standing about 18 in above the water surface at the highest point. The head was brown and snake-like, measuring approximately 1 ft across the top, and raised 18 in out of the water.

Moray the Monster of loch Morar

Representation of Loch Morar monster (© Michael Playfair)

Could certain Scottish lochs be home to gigantic eels or eel-like fishes? (© Richard Pullen)

Back in 1958, Morag gained a most unusual accolade, one that is unlikely to be claimed by any other lake monster. This was when scientist George Cooper included in a watercolour a large hump that suddenly appeared on the surface of Loch Morar while he was painting it.

On 1 August 1996 came the electrifying news that Cameron Turner, a diver from Darlington, had discovered some bones from a large unidentified animal at a depth of 60 ft in Loch Morar. Could these be the mortal remains of a Morag? Sadly, no - the following day a biologist formally identified them as the bones of a red deer *Cervus elaphus*, a common Highland species.

In April 2005, English cryptozoologist Richard Freeman, together with Lisa Dowley and David Curtis, spent some days at Loch Morar in the hope of enticing its cryptic monster Morag into view. Their bait was a mixture of mussels, fish guts, herring, cow liver, and a fish-attracting chemical called Van Den Eynd Predator Plus, placed inside permeable Hessian sacks, floating 20 ft or so beneath the water surface, attached to flotation devices. Richard believes that reports of monsters in various Scottish lochs, including Morar and Ness, may be due to the existence of extra-large, sterile eels nicknamed eunuch eels, which, instead of migrating to the sea to breed, remain permanently in freshwater where they continue to grow, attaining great sizes (see also Chapter 7). The reality of such eels, however, has not been scientifically confirmed. In any case, although the flotation devices were successful the bait itself was not touched.

Returning to the media headlines in 2013 after two decades of cryptozoological reticence, a fairly recent claimed encounter with Morag featured a trio of sightings in close succession. For within the space of just two days during summer 2013, holidaymakers Doug and Charlotte Christie from Brechin in Angus apparently saw the monster on three separate occasions while staying at Kisimuil bed-and-breakfast at the lochside. They saw a 20-ft-long black object in the middle of the loch, for 10 minutes on the longest occasion before it submerged again. Charlotte likened it to a whale, Doug to a submarine. In 2015, a further sighting was reported, this time by a local man who has been sailing the loch for some 20 years.

LOCH OICH
Wee Oichie or Oichy of Loch Oich, directly below Loch Ness and 4 miles long, traditionally sports a flattened head rather than the familiar equine form often noted for Nessie and various other Scottish loch monsters. Having said that, the head of the very big, black, serpentine beast that rose to the surface one summer's day in 1936 was vaguely dog-like, according to A.J. Robertson who spied it while boating at the loch's southwestern end. Certain other eyewitnesses, moreover, including a former loch keeper at Oich interviewed by investigator J.W. Herries during the 1930s, have likened Wee Oichie to a huge otter.

As a river connects Loch Oich to Loch Ness, some researchers have speculated that perhaps Wee Oichie and Nessie are one and the same, merely swimming back and forth from one loch to another via this interconnecting river. Indeed, during the mid-1930s, Herries interviewed three eyewitnesses who claimed to have actually observed such an animal journeying via this exact manner from Ness to Oich.

A more recent Oichy sighting occurred on 22 August 1998, when two Lochaber locals who wish to remain anonymous saw a large dark-coloured hump, rough but symmetrical in shape, break the surface a few hundred yards east of the Well of the Heads and about 22 yards from the shore as they were driving along the road next to the loch. Interestingly, they could see underneath the hump, thereby indicating that the creature was coiled and elongate. The two eyewitnesses got out of the car and ran onto the beach, armed with cameras, but the hump had already gone back down. Readily discounting identities such as swimming sheep, a line of otters, a seal, deer, and other commonly-posited candidates, they speculated that it might have been an eel, but with what they estimated to be a diameter of 18 in, if so it would have been one of truly prodigious proportions.

LOCH QUOICH

Situated west of Garry roughly 25 miles northwest of Fort William in Lochaber, Highlands, Loch Quoich is 9 miles long, with a maximum depth of 281 ft, and is supposedly home to a horse-headed but markedly serpentiform water monster. During the early 1930s, one such creature was even allegedly witnessed on land, when an unnamed lord, fishing on the loch's shores, spied it lying on a stony beach near to the water. It was also seen by the two fishing guides accompanying him, but he swore them to secrecy, afraid that the locals would consider all three of them inebriated, so their accounts remained unreleased for many years.

Creatures looking like this have been reported from Loch Quoich
(© Richard Svensson)

A collection of monster reports from Loch Quoich and other Scottish freshwater lochs was compiled by Father Henry Cyril Dieckhoff, from the Benedictine Abbey at Fort Augustus. Sadly, however, he died in 1970 before completing a book that he had been preparing, and which would have contained all of these reports.

LOCH SHIEL

Loch Morar is a famously remote lake, much of which can be reached only by boat, but this is also true of Loch Shiel - Scotland's fifth largest loch, with a length of 17 miles, a width ranging from 100 yards to a mile, and a maximum depth of 420 ft. Its own resident monster is known as Seileag.

Seileag's most diligent investigator was the afore-mentioned Father Henry Cyril Dieckhoff, who collected many reports. One of these, dating from 1905, featured Ewan MacIntosh, two young boys, and an old man called Ian Crookback, all of whom observed three humps above the water surface with the aid of a telescope while travelling across the loch opposite Gasgan aboard the little mail steamer *Clan Ranald*.

A massive creature with a broad head, wide mouth, long thick neck, and seven "sails" (humps) on its back was viewed through a telescope by Ronald MacLeod as it emerged from the water at Sandy Point one afternoon in 1926. Indeed, it was claimed by MacLeod to be bigger than the *Clan Ranald*!

LOCH TAY

Scotland's sixth largest loch by area and over 490 ft deep at its greatest depth, Loch Tay is situated in Perthshire, is approximately 14.5 miles long and typically 1.0-1.5 miles wide. Cryptozoologically speaking, however, the most notable mystery beast from this vicinity was not reported in the loch itself but from the nearby Firth of Tay, during the late evening of 30 September 1965. Moreover, it was actually seen on land, and therefore in full – a rare event indeed. Here is my documentation of that very remarkable incident, from my book *In Search of Prehistoric Survivors* (1995):

> [It] was brought to cryptozoological attention by veteran monster hunter F.W. Holiday. It was 11.30 pm, and Maureen Ford (wife of amateur flyweight boxer David Ford) was driving with some friends along the A85 by car towards Perth, in northeastern Scotland. Close to Perth, Ford suddenly spied an extraordinary creature by the roadside, only a few yards from the banks of the River Tay, which enters nearby into the Firth of Tay - an inlet of the North Sea. She described it as: "...a long grey shape. It had no legs but I'm sure I saw long pointed ears."
>
> Less than 2 hours later, it was seen again - but this time on the opposite side of the road, to where it had evidently crossed during the intervening period. At 1 am, Robert Swankie was driving along the A85 away from Perth towards Dundee, when his headlights revealed an amazing sight. As he later revealed in a Scottish *Daily Express* report (5 October 1965):
>
> "The head was more than two feet long. It seemed to have pointed ears. The body, which was about 20 feet long, was humped like a giant caterpillar. It was moving very slowly and made a noise like someone dragging a heavy weight through the grass."
>
> Swankie slowed down, and opened his window, but he could see another car not far behind, so he decided not to stop, and continued his journey. His testimony, and also that

of Ford, were taped by an enthusiastic investigator, Miss Russell-Fergusson of Clarach Recordings, Oban, and the police were also informed. In the *Express* report, one of their spokesmen commented that in the dark the headlights of a car could play tricks when they strike walls and trees - but as Holiday sensibly pointed out, if Swankie's sighting had merely been an optical illusion, why didn't he see monsters throughout his road journey? And how can an exclusively visual deception create a dragging sound?

Far more reasonable, surely, is the scenario of a reclusive sea creature emerging under the cover of darkness from the Firth of Tay, possibly via the River Tay itself, and, by sheer chance, being seen by two night-travelling eyewitnesses during its brief overland foray.

A popular cryptozoological identity for highly elongate water monsters is an evolved, modern-day species of zeuglodont whale, possessing a more flexible vertebral column than that of fossil forms and therefore capable of performing the vertical undulations often reported for serpentiform aquatic cryptids. Might this be what emerged from the Firth of Tay 50 years ago?

Reconstruction of a zeuglodont (© Tim Morris)

LOCH TREIG

A reservoir since 1929, the ominously-named Loch Treig (Scottish Gaelic for 'Lake of Death') is 5.6 miles long, and is located in a steep-sided glen just over 12 miles east of Fort William in Lochaber, Highlands. According to local medieval folklore, it was home to ferocious water-horses, but mystery beasts have also been reported here in modern times. Indeed, in 1933, during the creation of the extensive hydroelectric scheme now present in this area encompassing Treig, B.N. Peach, an engineer in charge of that scheme, stated that some of the divers working on the project had quit the job or had asked to be moved to other jobs because they claimed that there were monsters in this loch's depths.

LOCH WATTEN

Last – and definitely least – is Wattie, the infamous monster of Loch Watten, infamous inasmuch as its history owes precious little to cryptozoology, and even less to reality, as I discovered when conducting the only detailed investigation ever undertaken into this extraordinary case. All is revealed in Chapter 14 of this present book, which is devoted entirely to Wattie.

As with the Nessie saga, many sober sightings have been reported at these Scottish lochs that do appear to feature something more than misidentified otters, seals, sturgeons, birds, boats, algal mats, and suchlike - but what? All of the familiar cryptozoological Nessie contenders have been offered - a surviving plesiosaur, an undescribed species of long-necked seal, an elusive modern-day version of the officially long-extinct elongate zeuglodont whales, a giant form of eel - but with no physical evidence to examine, no firm identification can be offered.

If such reports as those documented here are indeed genuine, however, it seems likely that the species responsible can actively migrate overland, or via connecting rivers, from one loch to another (eels readily come to mind here). This would explain sightings in bodies of water that are too small or insufficiently stocked with fish and other potential prey to sustain a permanent, viable population.

THE MINI-MONSTERLINGS OF LOCH NESS

Finally: this chapter couldn't close without surveying the littlest lesser Nessies – or the most mini-monsterlings – of all, but to do that we must return to Loch Ness itself.

On 20 September 1981, an article penned by George Rosie on the subject of some very intriguing mystery beasts in miniature that had lately been filmed in Loch Ness appeared in no less august a publication than the *Sunday Times*. During July of that year, Mike Carrie and Jim Hogan had been scouring the loch bottom seeking the 29-year-old wreck of British racing driver John Cobb's jet speedboat, *Crusader*. (Tragically, Cobb had met his death on the loch after hitting at over 200 mph an unexplained wake - deemed by some to have actually been the LNM - while attempting to break the world water speed record on 29 September 1952 in *Crusader*, which had then sunk.)

They had been using a specialised underwater television camera that amplified light 2,500 times (essential in the peaty waters of Loch Ness) and was from Carrie's own company, Submersible Television Surveys Ltd. For some time, nothing but barren mud and rocks could be seen, then suddenly some small white life-forms flitted into view, which Carrie and Hogan videoed.

According to Carrie:

> I can best describe then as giant white tadpoles. They were about two or three inches long, white or pale grey in colour, seemed to have tails and swam just above the bottom.

Hogan, conversely, considered that they more closely resembled "wee white mice" with long tails and legs:

> They propelled themselves along the bottom in a jerky way. We knew what size they were because there was an extension to the camera to measure them against.

Greatly intrigued by these highly unexpected and very baffling yet seemingly uncommon mini-monsterlings (they only saw a handful during 3 weeks of filming), Carrie and Hogan submitted their videotape of them to Dr P. Humphrey Greenwood, an ichthyologist at London's Natural History Museum (NHM), for his opinion as to what they may be. Equally fascinated, Dr Greenwood arranged for some computer-enhanced outlines of the mystery beasties to be prepared, which in turn revealed that they seemed to have three pairs of limbs or limb-like protuberances.

In a letter dated 18 April 1985 sent to veteran Lake Champlain monster investigator Joseph W. Zarzynski, who had written to him enquiring about this mystifying species, Greenwood stated:

> On the subject of that object I am certainly prepared to say that the information on its size, and from what I could determine of its form and locomotion in the videotape, I would suggest that it was a small crustacean, and certainly it was no vertebrate animal that I could identify as being part of the fauna of Loch Ness.

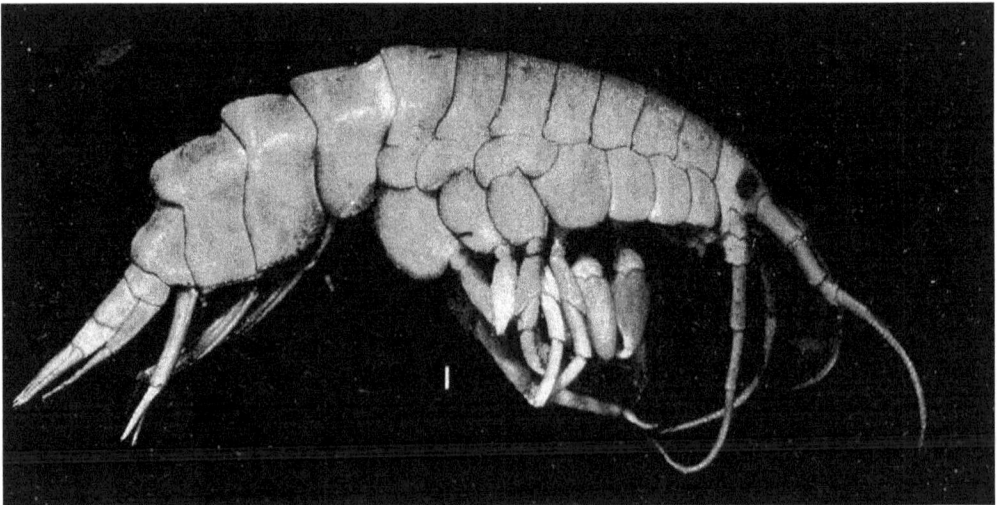

An amphipod (public domain)

Zarzynski reproduced the latter quote in a short chapter devoted to these entities (and also to *Tullimonstrum* – see Chapter 7 of this present book of mine) in his book *Monster Wrecks from Loch Ness and Lake Champlain* (1986).

Additionally, Rose had noted that Greenwood's best guess was that (in Rose's words): "it could be some kind of bottom-dwelling crustacean, hitherto unknown in Loch Ness, similar to some found in other deep-water lakes such as Lake Baikal in Siberia".

When I read Rose's account in the *Sunday Times*, I thought straight away of amphipods, which are usually very small, superficially shrimp-like crustaceans represented by both marine and freshwater species. Some of these do have long tails and are often pale in colour, but they sport more than three pairs of legs (though some are only very small and slender, and thus may not have been rendered visible even by the computer-enhancement techniques that the NHM applied to the Carrie-Hogan videotape). So could these loch-bottom mini-monsterlings constitute a new amphipod species? Quite possibly - but without a specimen to examine, their taxonomic status presently remains unresolved.

Interestingly, however, the latter videotape was not the first evidence for the existence of Loch Ness's 'white mice' (the name by which these still-unidentified life-forms are most commonly known nowadays), or at least something like them. As far back as 1972, at around the same time that they obtained their famous underwater 'flipper' photographs (see Chapter 1) in the loch, longstanding Nessie seeker Dr Robert Rines and his team from the Academy of Applied Science (AAS) also filmed what may be the same 'white mouse' species, or possibly a smaller, related species, or even a juvenile version.

A photographic still from that film, which appeared in the AAS's 1972 publication, *Underwater Search at Loch Ness*, depicted a pale mystery monsterling that was somewhat bumblebee-like in shape (hence Rines and his team nicknamed such creatures 'bumblebees'). However, it sported a pair of long appendages stretched out horizontally that made it look surprisingly similar to those familiar surface-dwelling freshwater hemipteran insects known as water boatmen (genus

Basic outline of a Loch Ness 'bumblebee', based upon the image of one in a photographic still from the 1972 AAS film (© Dr Karl Shuker)

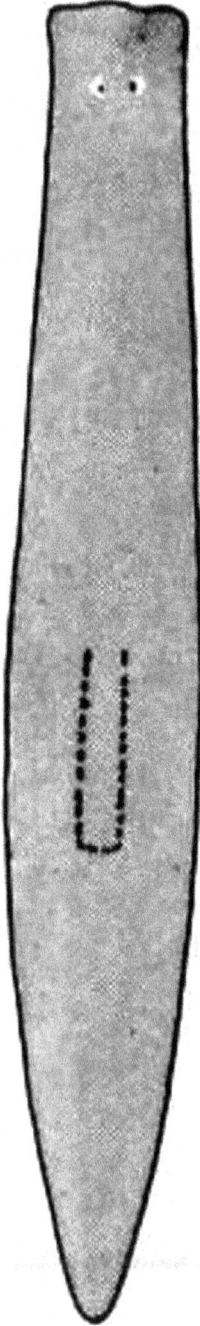

Corixa) and backswimmers (genus *Notonecta*). I am greatly indebted to American lake monster researcher Scott Mardis for kindly sharing with me several additional stills from the AAS's 'bumblebee' film, made available in turn to him by LNM investigator Dick Raynor, which corroborate this mystery creature's morphology as seen in the photograph contained in the AAS's earlier-noted 1972 publication.

Scott has likened the 'bumblebee' form to a veliger (the larva of gastropod molluscs), whereas Canadian cryptozoologist Sebastian Wang has compared it with tiny shelled crustaceans known as ostracods, and Loch Ness veteran Adrian Shine has suggested a cladoceran crustacean (aka water flea) from the genus *Bythotrephes*. Some mystery beast investigators have even speculated that perhaps it is a larval form of Nessie itself, but unless the latter is an invertebrate, as these 'white mice' and 'bumblebees' must surely be, this notion is untenable.

However, such a huge body mass as possessed by the adult Nessie (judging at least from eyewitness accounts) would surely require an internal skeleton in order to support it and give it shape, which thereby argues against an invertebrate identity for the LNM. This telling point was noted by American college student Jay Cooney in an interesting article dealing with Loch Ness's monsterlings, which he uploaded onto his *Bizarre Zoology* blog on 30 March 2014. (Unfortunately, however, this particular article later developed image-corruption problems, so Jay subsequently removed it from his blog.)

'White mice' and 'bumblebees' are not the only unexplained life forms discovered in Loch Ness. Once again during 1981, but this time in April, via a scientific paper written by Dr T.B. Reynoldson and two fellow University College of North Wales zoologists, and published in the *Journal of Zoology*, the remarkable news was made public that a creature which could legitimately be described as an alien worm had been discovered here too, and in some numbers. More specifically, it was *Phagocata woodworthi*, a species of North American triclad turbellarian flatworm not native to Europe and never previously recorded from anywhere outside the New World.

Up to 1.2 in long and 0.2 in wide, dark grey, brown, or almost black dorsally, paler ventrally, dorsoventrally flattened, and sporting a truncate head, this out-of-place (o-o-p) invertebrate (o-o-p species are

A species of *Phagocata* flatworm (public domain)

often referred to loosely as alien species) is known to attach its cocoons with limpet-like efficiency to the bottoms of maritime vessels. However, the likelihood that these could remain in situ during an entire transatlantic crossing seems unlikely – plus, few vessels traversing the Caledonian Canal have come from America anyway. So how did this species reach Loch Ness?

With ultimate irony, researchers have concluded that the likeliest sources of these worms are none other than the various monster-hunting vessels that have been transported down through the years from North America directly to the loch, in particular a mini-submarine imported here in 1977.

They do say that it takes all sorts to make a world, and certainly, from traditional water-horses, water-bulls, and turtle-pigs to modern-day long-necked, serpentiform, vermiform, and even bumblebee-like aquatic mystery beasts, this maxim is also clearly applicable to the cryptozoological world, at least as far as the multifarious monsters reported from mainland Scotland's freshwater lochs are concerned.

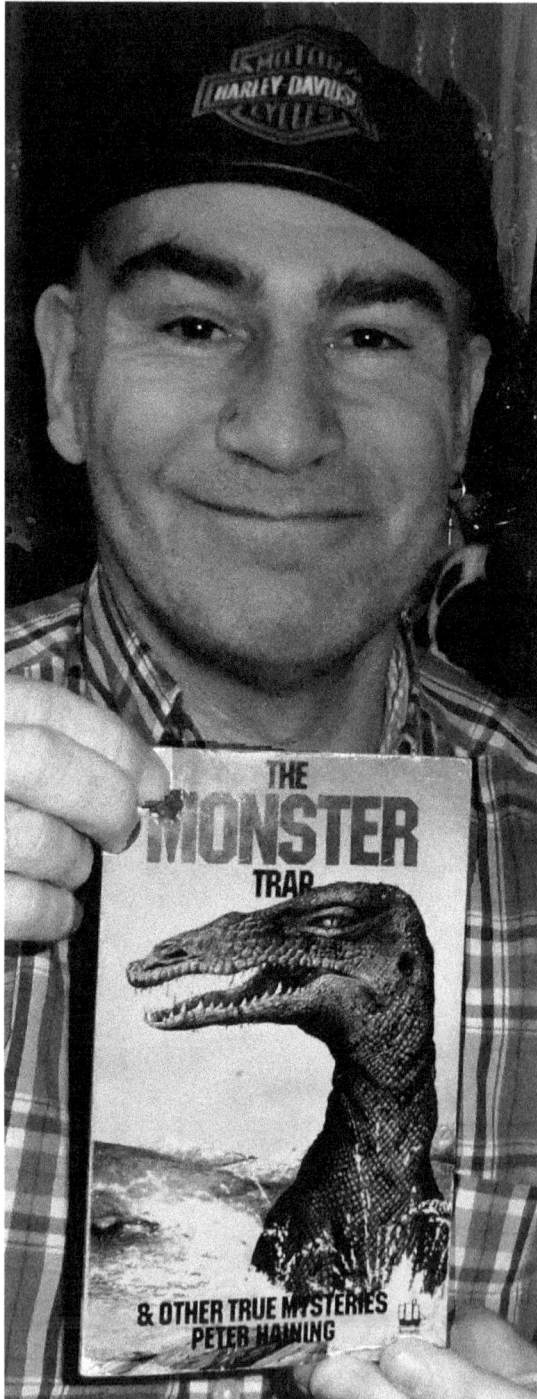

With my copy of *The Monster Trap*
(© Dr Karl Shuker)

CHAPTER 14:
WHATEVER HAPPENED T⊕ WATTIE? SEEKING L⊕CH WATTEN'S ⅢISSING Ⅲ⊕NSTER

> It was a bright spring day, and steely sunshine glinted over the mountains of Caithness when Colonel Arthur Trimble first saw the monster of Loch Watten.
>
> The monster's eyes were slits in a huge squat head, and its body, which loomed under the rippling water, appeared at least 20 ft wide. It observed him for several seconds. He even had time to take a photograph of it.
>
> John Macklin – *Leicester Mercury* (Leicester), 28 March 1966

I first learnt about this chapter's very intriguing subject when, as a teenager, I read a fascinating little book by prolific author Peter Haining entitled *The Monster Trap and Other True Mysteries* (1976). That same book also introduced me to various other equally thought-provoking subjects that I subsequently investigated in depth – winged cats, the Green Children, and the mysterious mini-mummy of Wyoming.

However, the chapter that interested me most of all (and which gave its title to the entire book) was also the one that has mystified me most of all – because, for over 30 years, and in spite of the fact that it was potentially of immense cryptozoological significance, its subject resisted every attempt made by me to uncover any additional details regarding it. Only in recent times did I finally unlock the secret to this highly perplexing matter, as now revealed.

THE (VERY) MYSTERIOUS CASE OF THE MONSTER TRAP
The setting for the truly extraordinary episode documented in this chapter is Loch Watten – a Scottish freshwater lake in Caithness's River Wick drainage system. Its grim tale as given in Haining's book (in which the chapters' stories, although all allegedly true, are written up in a dramatised, novel-like style) can be summarised as follows.

According to Haining, the incident in question took place some 10 years before the flap of Nessie sightings in 1933, and featured local estate owner Colonel Arthur Trimble (who had retired in 1922 from the British army). It all began on the morning of 21 April 1923, when Trimble was walking his spaniel, Bruce, by the lochside, not far from his estate. He had a camera with him, as it was a pleasant morning and he hoped to take some photographs. After reaching his usual point for turning back, Trimble called to Bruce, who had run some distance further ahead, and after waiting for him to come back, Trimble looked out across the loch – where, in Haining's words:

Something dark and looming had suddenly appeared on the surface of the water.

The Colonel squinted his eyes and raised his hand to half-shade his face. The form was clearer now. It looked like a kind of neck with a huge flat head.

Keeping quite still, he looked harder and could see that it was indeed a head and neck, and that there were slit eyes staring directly at him. Below the surface of the water he could make out the shape of an immense body, at least twenty feet wide.

Colonel Trimble could hardly believe the evidence of his senses. It seemed like some huge water monster.

Haining stated that although the monster was less than a hundred yards away, thanks to his years of army discipline Trimble did not panic, and lifted his camera. Just as he was about to take a photograph, however, his dog Bruce spied the monster and immediately ran towards it, barking loudly. Startled, the monster disappeared beneath the water almost at once, but at that same instant Trimble succeeded in snapping a single photo, although he had no idea whether he had actually captured the beast's image. When he and Bruce arrived back home and he told his housekeeper, a local woman called Mrs Doris Dougal, what had happened, she confirmed that he had seen the loch's legendary 'serpent', and suggested that he report his sighting.

That same day, Trimble took his camera's film to the local chemist shop for developing, and when he collected his photos two days later he was delighted to discover that although the picture snapped by him at the loch was slightly blurred, it did indeed depict the monster's head and neck above the water surface. Consequently, that afternoon he penned an account of his sighting for London's *Times* newspaper, enclosed with it a copy of his photograph, and posted it a few hours later. From then on, Trimble visited the loch daily, in the hope of seeing and photographing the monster again, but leaving Bruce at home to ensure that he didn't cause any disturbance this time if the monster should reappear.

Unbeknownst to Trimble, however, on 1 May, while he was once again at the loch, Bruce managed to sneak out, and when Trimble returned home later that day he was met by Mrs Dougal with the disturbing news that Bruce was missing. The two of them spent some time searching for the dog locally, but to no avail – until Trimble saw a man approaching from the direction of the loch. The man was Trimble's nearest neighbour, the local doctor Robert McArdish, who told Trimble that he had spied Bruce swimming in the loch – but just as the doctor had been about to call out to him, he had seen a flurry in the water, as if something else was also there, and then the dog disappeared, after which the waves settled again, but with no sign of Bruce.

Enraged by the apparent killing and devouring of his dog by the monster, the following day Trimble set about building an extraordinary 'monster trap', consisting of 50 fathoms of rope attached to an enormous sharpened spike of steel that had been shaped into a massive hook. Trimble baited this hook with a large piece of freshly-purchased horsemeat, and after rowing into the middle of the loch in his dinghy he lowered the hook into the water, attached a marker buoy to the end of the rope, and dropped it overboard. Then he rowed back to shore, and returned home.

The next morning Trimble went out to inspect the trap, but it had not been touched, so he repositioned it elsewhere in the loch, and came back home. This procedure was repeated up until the evening of 4 May, when he informed Mrs Dougal that he was going out to the loch again, even though it was almost dark. Just on 9.30 pm, after looking outside to see whether he was returning as he was late, Mrs Dougal suddenly heard a single loud, terrified scream, from the direction of the loch. Racing outside to the gardener's cottage close by, she hammered on his door, explained what had happened, and the two of them ran fearfully to the loch. There, in some reeds at the lochside, was the half-submerged body of Trimble, and as they looked down at it, they saw to their horror that his chest had been pierced by the giant hook, which was still attached to the rope. And as they stood there, they heard something:

...something that turned their blood to ice – and haunted them for the rest of their days.

It was a sound which came from the loch. The sound of something large that splashed as it swam away from the shore...

And with that dramatic little flourish, there endeth Haining's tale of the Loch Watten monster (let's call it Wattie, for short).

WHITHER WATTIE?
Needless to say, one would imagine that such an episode, far more sensational than anything that even Nessie can lay claim to, would have subsequently featured in every major (and minor!) cryptozoology publication as a matter of course, as famous – or infamous – as the story of the Surgeon's Photograph and other endlessly rehashed and recycled cryptozoological histories. Yet nothing could be further from the truth.

Indeed, when the original version of this ongoing investigation of mine was published by *Fortean Times* in its September 2009 issue, I had yet to discover a single mention of the Wattie history anywhere – I knew of no book, periodical paper, magazine article, newspaper report, or website that contained even the briefest reference to it. Moreover, the only acknowledged claim to fame of Loch Watten, other than having been formally designated as an SAC (Special Area of Conservation), is that it is a good body of water for fly-fishing for brown trout. In stark contrast, any celebrity status as a monster-haunted lake is conspicuous only by its absence. So how can such anomalies be explained?

Let's look at some background information, beginning with a few additional details supplied by Haining himself. In his book's introduction, he stated that when selecting stories to be covered by him in it, he didn't want to repeat ones that lots of other writers had already utilised. Instead, he decided:

> I would use stories that had particularly fascinated me in which I had done considerable research, if not actually visited the places in question.

Furthermore, in the opening to the 'Monster Trap' chapter itself, he stated that although Nessie was certainly the most famous Scottish monster, she was not the only one, noting that there were stories of water-horses and serpents from many other Highland lochs, and then commenting:

> One particular monster story has always fascinated me, but amid all the fuss about 'Nessie' it rarely gets mentioned.

Loch Watten (public domain)

For 'rarely', substitute 'never'!

Yet according to Haining's book, local people claim that there have been stories of a monster, which they term 'the serpent', in Loch Watten for many years, but no documented records of actual sightings prior to Trimble's ultimately fatal incident. Is this true? Never having visited the loch myself, which is only 14 miles from John O'Groats in the far northeast of Scotland, I have no idea whether there is any verbal tradition of a monster here (though I have yet to communicate with anyone versed in Scottish mythology or cryptozoology who has ever heard of such tales). However, I would have expected at least some documentation of it, were such a tradition to exist. After all, as Haining correctly pointed out, there are accounts of monsters for a number of other lochs – including Ness, Morar, Oich, Lochy, Shiel, Arkaig, Lomond, Quoich, and Treig (see Chapter 13 in this present book of mine).

Another anomaly concerns Loch Watten itself. Despite being the second largest of Caithness's lochs, it is under three miles long, less than a mile across at its widest point, and boasts an average depth of only 10-12 ft – a very far cry from the immeasurably greater size of Loch Ness, Loch Morar, and other notable bodies of Scottish freshwater associated with monster traditions. If the kind of huge reptilian monster (at least 20 ft *wide* – so how *long* was it?!) allegedly encountered by Trimble were truly real, it would surely require a much more substantial aquatic domain than Watten.

Nor do these inconsistencies constitute the full extent of my concern for the validity of Wattie as a bona fide cryptid. When I first attempted to research this subject, back in the early 1990s, I wrote on two separate occasions to Haining, having obtained his correct address, but I never received a reply to my requests for information, and as he died in 2007 this most direct line of investigation is no longer an option. In addition, I met with a succession of dead-ends when attempting to uncover any Trimble-related leads (not even trawling through death registers and army records online elicited any evidence for his supposed former existence). I also searched meticulously through the relevant period of back issues for *The Times*, but did not find any published letter or photo by Trimble.

In short, prior to the publication of my *Fortean Times* article, the only known source of information (to me, at least) concerning Wattie was Haining's book, and, therefore, Haining himself – which to my mind is the most disturbing aspect of all concerning this mystifying tale. The reason why I say this is that some of Haining's other publications have already attracted considerable controversy in relation to the validity – or otherwise – of their claims.

SPRING-HEELED JACK, SWEENEY TODD, AND WOOLPIT'S GREEN CHILDREN – THREE REASONS FOR WATTIE WORRIES

For example: in a detailed paper on Spring Heeled Jack (*Fortean Studies*, vol. 3, 1996), Mike Dash revealed that he was unable to obtain independent corroboration of various accounts and details that had been published by Haining in his book on this subject (*The Legend and Bizarre Crimes of Spring Heeled Jack*, 1977). And even an engraving claimed by Haining to show the recovery from a marsh of one of Jack's victims – a victim, incidentally, undocumented by anyone else – in reality showed no such thing.

Moreover, when Mike Dash wrote to him asking for sources, Haining replied that he was unable to supply any because all of his research material had been loaned to a film scriptwriter who had subsequently vanished. Not surprisingly, perhaps, in his paper's annotated bibliography, Mike made the following comments regarding Haining's book:

> The only full-length work on the subject is a curious hodge-podge of the accurate, the overtly-dramatised and the invented...it repeats many existing errors, creates new ones, and is so single-mindedly determined to fit evidence to the theory that Jack was the Marquis of Waterford that it does not flinch from introducing made-up evidence to support this case.

Equally controversial are Haining's books on Sweeney Todd, Fleet Street's homicidal hair-snipper. Although Todd is widely assumed to be an entirely fictitious character spawned by the Penny Dreadfuls of Victorian times, Haining published two book-length treatments, respectively entitled *The Mystery and Horrible Murders of Sweeney Todd, the Demon Barber of Fleet Street* (1979) and *Sweeney Todd: The Real Story of the Demon Barber of Fleet Street* (1993), in which he alleged that such a person had actually existed. However, this claim has attracted much criticism, for a variety of reasons, including those summarised succinctly in Wikipedia's entry for Haining (as accessed by me on 2 July 2009):

> In two controversial books, Haining argued that Sweeney Todd was a real historical figure who committed his crimes around 1800, was tried in December 1801, and was hanged in January 1802. However, other researchers who have tried to verify his citations find nothing in these sources to back Haining's claims. A check of the website 'Old Bailey' for "Associated Records 1674-1834" for an alleged trial in December 1801 and hanging of Sweeney Todd for January 1802 show no reference; in fact the only murder trial for this period is that of a Governor/Lt Col. Joseph Wall who was hanged 28 January 1802 for killing a Benjamin Armstrong 10 July 1782 in "Goree" Africa and the discharge of a Humphrey White in January 1802.

In short, there are notable precedents when faced with questioning the reliability of claims made by Haining in the absence of any independent sources of evidence to examine. Even in another chapter of *The Monster Trap,* documenting the Green Children, it is curious to note that the famous, historically-recorded incident of the Woolpit Green Children receives no mention whatsoever. Instead, Haining devotes the entire chapter to an exceptionally similar version allegedly occurring several centuries later in Spain – a version subsequently revealed by other researchers to be a complete fabrication, by person(s) unknown, directly inspired by the Woolpit episode.

And I hardly need point out that Haining's description of Trimble's supposed photo – slightly blurred but showing a head and neck – is more than a little reminiscent of the Surgeon's Photograph of Nessie. Also worth remembering is that aside from his non-fiction books, Haining was a well-respected, extremely knowledgeable anthologist of short stories of fiction in the horror and mystery genres.

FOUR WATTIE UPDATES

The publication by *Fortean Times* in its September 2009-dated issue (but on sale in August) of my original Wattie article that forms the basis of this present chapter subsequently triggered four notable responses.

The first of these was an email of 28 August 2009 that I received from Rod Williams of Talgarth, Wales:

> I am a regular reader of *Fortean Times* and your item on Wattie was interesting but feel that it was a concocted tale by Peter Haining.
>
> I have read Hugh Miller's *Scenes and Legends of the North of Scotland* and also Samuel Smiles's biography of Robert Dick [*Robert Dick, Baker of Thurso, Geologist and Botanist*, 1878], baker, biologist (botanist mostly) and geologist. A man who walked many miles at night over large parts of Caithness.
>
> I cannot recollect in either book mention of Loch Watten or a/its monster. Both men were not above mentioning curious tales, particularly Miller who was well into hauntings and weird happenings; apart from being a quarryman turned geologist he seemed to thrive on such tales.
>
> I may have missed any reference of course but the book of Miller's can be read on line for free should you wish to check it out.
>
> Not sure of Dick's biography being on line but probably is.

**What were the 'crocodiles' said by George Borrow
to inhabit certain Welsh lakes?**

George Borrow's *Wild Wales* (circa 1854) mentions 'crocodiles' in Welsh lakes or rather stories of these mythical beasts and enquires of people on his journey whether they knew of any local legends relating to these little lakes and crocs.

Again I don't remember any specific stories as it has been many years since reading the book. I think I need to re-read it sometime.

Our local lake Llangorse Lake has large pike in it and one chap told me that when he was wind surfing and was stood in the lake (shallow in many places) something large brushed his leg.

Quite apart from confirming the absence of Wattie information from some literary sources new to me, Rod's email is also of value for the interesting snippets of information concerning Wales's mystifying water 'crocodiles', which I've read about in a number of publications and which deserve a detailed examination in their own right.

The second response was a letter penned by German cryptozoologist Ulrich Magin, which was published by *Fortean Times* in November 2009. In his letter, Ulrich revealed that Haining's account of Wattie was almost identical to a tale included by French fiction writer George Langelaan in *Les Faits Maudits* (not *Maufits* – as erroneously titled in Ulrich's published letter) or 'Cursed Facts' - a book of forteana published in 1967, containing an eclectic mixture of retold press clippings and fictional stories. Langelaan claimed that his source for that particular tale was a *Times* news report from May 1932, but a search for it undertaken by Ulrich failed to unearth any such report.

The third, highly significant response was a letter that was forwarded to me by *Fortean Times* on 31 March 2010, which had been written to me on 23 March by Lance Shirley of Cornwall and was accompanied by a remarkable enclosure – a photocopy of an article that had been published in the *Leicester Mercury* newspaper on 28 March 1966 in what appears to have been a regular, long-running series of articles published under a 'Stranger Than Fiction' banner. Written by a John Macklin, the article was entitled 'The trap he set was for a monster...but it was the colonel who died'. Reading it through, I discovered that its content and wording were so similar to Peter Haining's chapter 'The Monster Trap' that it seemed highly likely either that Haining had directly copied Macklin's account or that he and Macklin were one and the same person.

As I learnt from Mike Dash during correspondence with him of 1 and 2 April 2010, Haining is known to have written under various pen-names as well as his own, so could John Macklin be yet another one? After receiving Lance's letter and enclosure, I googled John Macklin on the internet, and discovered that just like Haining, he is/was a prolific author, and, again just like Haining, has authored many popular-format compilation books of supposedly true mysteries. Just another coincidence?

In his letter to me, Lance mentioned that he and his family had lived in Caithness, near to Loch Watten, from 1966 to 1976, during his childhood. While still living there in the early 1970s, he had read the *Leicester Mercury* article, which had belonged to his mother (it had been forwarded to her for its interest value from her father, who lived in Loughborough and always bought this newspaper), and was excited to think that such a creature may live so close to them. Whenever they passed the loch in the car, they always scanned the surface, just in case they could catch sight of the monster. Upon reading my *Fortean Times* Wattie article of September 2009, Lance realised that Haining's account matched what he could still recall from that newspaper cutting from long ago. Moreover, while subsequently clearing out the loft in the family home, he was delighted to discover it, yellowed with age but still intact, stored inside a biscuit tin crammed with other cuttings (including another John Macklin 'Stranger Than Fiction' article from the *Leicester Mercury*, this time dating from 1969 and documenting a ghostly occurrence in Hoy Harbour).

As Lance pointed out, what is so interesting is that the *Leicester Mercury* article predates not only Haining's book (by 10 years) but also that of George Langelaan (by a year). Consequently, it now seems that Langelaan did not originate this tale after all. Regardless of who did do so, however, no independent, substantiating evidence for its veracity or the existence in Loch Watten of a mysterious creature has ever come forward. Consequently, in my opinion the most reasonable conclusion remains that Wattie is a complete invention.

Irrespective of this, after receiving Lance's letter I lost no time in pursuing the Macklin line of investigation further. My ultimate goal was the procurement of some current contact details if he were still alive (in which case, of course, he and Haining could not be the same person!); or, if he were dead, uncovering as much biographical information concerning him as possible, in the hope of determining conclusively whether or not John Macklin was indeed merely another pen-name of Peter Haining.

In April 2010, I emailed an enquiry to Macklin regarding Wattie via Sterling, the American publisher of the most recent Macklin book that I have yet been able to trace (a children's book of true ghost stories, published by Sterling in 2006). So far, however, I have yet to receive any response from him.

Moreover, to me it seemed undeniably thought-provoking that whereas Macklin and Haining are/were both extremely prolific authors who wrote on extremely similar subjects, none of Macklin's works are cited in the bibliographies of any of Haining's books accessed by me (or vice-versa). Equally, I have been unable to trace any indication that Macklin has published any books or articles in the years following Haining's death. And whereas photos of Haining are readily obtainable by googling his name, Google is currently (as of March 2016) unable to locate a single photograph of Macklin. Also, whereas Haining has a detailed entry in Wikipedia, Macklin (despite being a comparably prolific – and hence successful - author) has no entry whatsoever.

So were Haining and Macklin the same person, with Wattie merely the figment of an inordinately prolific writer's fertile imagination? I was soon to discover the answer, which provided yet another unexpected surprise.

On 8 May 2010, I received the fourth response to my enquiry for Wattie information. This time it was a highly illuminating email from none other than *Fortean Times*' own co-founder, Paul Sieveking, who informed me that John Macklin was indeed a pseudonym – but not of Peter Haining! Instead, it was one of many pen-names used by another author of popular-format writings on mysteries – Tony James. The plot thickens! So did Tony James originate the storyline for the Wattie tale, or is there an even earlier version out there somewhere that he had read? If anyone has current contact information for James, I'd very much to hear from you!

Meanwhile, my sincere thanks go to Rod Williams, Ulrich Magin, and especially Lance Shirley and Paul Sieveking for shining some important new light upon this increasingly complicated mystery, and I am intrigued to see whether any further developments will occur in the future. After all, as a certain cult television series used to proclaim, the truth is out there – it's finding it that's the problem!

WAS THERE EVER A WATTIE? OVER TO YOU!

It gives me no pleasure whatsoever in questioning the legitimacy of the Wattie affair as documented by Haining, especially as the book in which it appears is one that has been instrumental in introducing to me various other subjects that have since become significant in my own researches – and I would therefore be delighted if my concerns regarding this case could be convincingly dismissed. Yet it is clear that the omens for Wattie's validity are not good at all.

Nevertheless, it would be rash to deny this tantalising tale out of hand without at least waiting to see whether this present chapter's re-dissemination of its currently-available information results in any additional findings coming to light. So here, gentle readers, is where you come in. If there is indeed anyone out there with direct or indirect, integral or background information relating in any way to monsters reported from Loch Watten, and to the Trimble incident in particular, I'd love to hear from you.

Similarly, if Haining's research files have been preserved, any details of where and whether they can be accessed would be very welcome. After all, if we are to believe his claim that all of the subjects in his book were ones in relation to which he had conducted considerable research, these archives undoubtedly offer the most likely source of primary and additional data concerning this most monstrous of Scottish crypto-mysteries.

CHAPTER 15:

PARAD⊕X ⊕F THE PICTISH BEAST

Occurring some fifty-four times, the 'beast' or 'Pictish elephant' symbol is perhaps the most enigmatic of the anamorphic symbols to appear on Pictish stones. Despite representing no known animal, its appearance on different stones across the country is surprisingly consistent. It appears to have a trunk-like snout, a tail, and a crest on its head. Its feet, however, are distinctly odd, ending in almost ethereal spiral structures.

Iain Forbes - *The Last of the Druids: The Mystery of the Pictish Symbol Stones*

Named after their very distinctive body tattoos, the Picts ('painted people') inhabited northeastern Scotland as a separate tribe from c.300 AD to 850 AD, after which they were united with the Celtic Scots under the reign of King Kenneth I.

The Picts can boast as their principal claim to archaeological fame their ornately-carved symbol stones. These are elaborately decorated with various creatures, objects, and other depictions, especially the earlier, pre-Christian stones - which are designated as Class I (dating from the 6th Century, generally unshaped, and bearing line-incised symbols on at least one flat face) or Class II (of rather later date, and bearing much more intricate, flamboyant designs). Class III stones, conversely, date from when Christianity reached the Picts, so on these stones the earlier Pictish symbols have been mostly replaced by Christian ones.

Due to their realistic designs, the many different animal types carved on Class I and II Pictish symbol stones are readily identifiable – with one notable exception, that is. Appearing on over 50 Class I and Class II stones, this bizarre-looking exception is known as the Pictish beast.

Line drawing of the Pictish beast (public domain)

219

Several very famous Pictish symbol stones bear depictions of it. These include: the Dunfallandy Stone (Class II) in Tayside; one of the Rhynie Pict stones in Aberdeenshire; and the 6-ft-tall Rodney's Stone (Class II), which is a cross-slab of grey sandstone originally present in the graveyard of the old church of Dyke and Moy but subsequently transferred to the Grampian village of Dyke to commemorate Admiral Rodney's victory and standing today on the left side of the avenue leading to Brodie Castle. Other symbol stones depicting the Pictish Beast are a cross-slab on the Brough of Birsay at the northwestern corner of Mainland, Orkney; the 9th-Century, 10-ft-tall Maiden Stone near Pitcaple in Aberdeenshire; and a carved stone in Grampian's Port Elphinstone Henge near Inverurie (the henge itself is much older than the carvings). Perhaps the least stylised, most 'natural' portrayal of this mystifying creature can be found upon a spectacular Class II stone at Tayside's Meigle Sculptures Stone Museum. It is adorned with carvings of horse riders and a tail-biting serpent as well as the Pictish beast, plus the customary Pictish V-rod and crescent symbols.

In September 2011, Scottish archaeologist Cait McCullagh, co-founder and director of Archaeology for Communities in the Highlands (Arch), publicly announced that an archaeological team had recorded a Pictish symbol stone built into the wall of a Highland farm building in the Black Isle, and that one of its three symbols was a Pictish beast. It is

Pictish beast depicted on Meigle 4 Stone at Meigle Sculptures Stone Museum
(© Simon Burchill/Wikipedia http://creativecommons.org/licenses/by-sa/3.0/)

mystifying why this stone had not been officially recorded earlier, and McCullagh suggested that more Pictish symbol stones may still await documentation in this locality, even though it was unusual to find such carvings on the north side of the Moray Firth. Intriguingly, the Pictish beast carving showed a lack of weathering, indicating that the stone may have been kept inside, or had been buried, for a long time before being placed in the byre's wall.

Depictions of the Pictish beast on such symbol stones as those noted here portray this bizarre creature with a dolphin-like head, a long beak, four limbs that often curl backwards underneath its body (although sometimes, as on the Meigle Museum stone, only the paws curl backwards), an elongate tail with a noticeable curl at its tip, and, most distinctive of all, what may be a long slender horn, crest, or even a trunk-like projection sprouting from the top of its head and curving over its back. Indeed, this last-mentioned feature has earned the Pictish beast the alternative name of 'swimming elephant' (which all too readily conjures up some decidedly surreal images of a Celtic Trunko!).

Needless to say, no known species of animal resembles the Pictish beast as so portrayed, which in turn has incited appreciable debate and controversy among historians and archaeologists as to what it may be. One popular, conservative identity for it is a dolphin, or even a beaked whale (ziphiid), based upon its beaked, superficially cetacean-like head.

Leading on from this particular line of speculation, I wonder if its anomalous 'trunk' may in reality be a representation of a spout of water spurting upwards when the dolphin or ziphiid exhales through its blowhole (conjoined, modified nostrils), which is indeed situated on the top of these marine mammals' head. Conversely, the unequivocally leg-like limbs and non-fluked tail of the Pictish beast are radically different from the flippers and fluked tail of all modern-day cetaceans.

Other postulated suggestions include a seahorse (especially when depicted vertically), a deer, a seal, and a dragon. A bona fide elephant or even an unknown species of secondarily aquatic elephant has also been considered (albeit not seriously, for obvious reasons!). It may simply be that the Pictish beast is an entirely fictitious, imaginary creature,

Close-up of Pictish beast depicted on Meigle 4 Stone at Meigle Sculptures Stone Museum (© Simon Burchill/Wikipedia http://creativecommons.org/licenses/by-sa/3.0/)

possibly even a composite of several different beasts, but its numerous portrayals (accounting for approximately 40 per cent of all Pictish depictions of animals) imply that it had considerable symbolic significance for the Picts.

Indeed, it may even be the earliest known artistic representation of the legendary kelpie or Scottish water-horse (see Chapter 12). One of the three Aberlemno symbol stones in Tayside depicts a pair of interlaced horse-headed, elongate aquatic monsters, and some scholars have suggested that these may constitute a more sophisticated version of the Pictish beast.

Is the Pictish beast an early depiction of mythological kelpies or water-horses? (public domain)

Moreover, in their book *Ancient Mysteries of Britain* (1986), Janet and Colin Bord proposed that the Pictish beast might be a direct representation of the elusive water monsters allegedly inhabiting various of Scotland's lochs, its 'trunk' explaining the familiar 'head and neck' or 'periscope' images often reported and even photographed by Nessie eyewitnesses. Backing up their fascinating hypothesis, the Bords make the following very telling observation:

Since a whole range of animals and birds is accurately depicted on the symbol stones - wolf, bull, cow, stag, horse, eagle, goose - perhaps these were the creatures most familiar to the Picts in their everyday world, and 'monsters' were also familiar to them, being more often seen in the lakes than they are today, and accepted as part of the natural world just like eagles and stags.

This in turn leads to the most intriguing and original (if zoologically offbeat) identity ever put forward for the Pictish beast. A familiar figure in the British fortean community for many years, Tony 'Doc' Shiels describes himself as a monster-hunter, stage magician, surrealist artist, and shaman of the western world (among other things), and he has suggested that the Pictish beast may indeed be a depiction of the unidentified Scottish water monsters. Moreover, as he first documented in a *Fortean Times* article (autumn 1984) and further propounded six years later in his book *Monstrum! A Wizard's Tale* (1990), and as I have also referred to briefly earlier in this present book (see Chapter 7), he has speculated that these latter mystery beasts' zoological identity could in turn be a highly novel, specialised form of squid.

But how could such a creature be equated with Nessie and company, and how firm are its basic anatomical and physiological foundations? Here is what I wrote about Shiels's proposed 'Pictish squid' in my book *Mysteries of Planet Earth* (1999):

> As conceived by Shiels, the most striking feature of his hypothetical species is a long, flexible, prey-capturing proboscis-like structure (the trunk of the Pictish beast), on account of which he has dubbed this creature the elephant squid. If held out of the water, its proboscis could resemble a long neck, which Shiels believes may explain the familiar 'long-neck' images of Nessie and her kin. He also provides his elephant squid with inflatable dorsal airsacs as part of its buoyancy mechanism (which could yield the varying shape and number of humps reported for Nessie), six short tentacles, and a pair of longer curling arms (the Pictish beast's curling front legs), as well as a muscular tail bearing two horizontal lobes.
>
> In his accounts, Shiels proposes that this remarkable mollusc may even be able to emerge briefly onto land, which might therefore explain why certain Nessie eyewitnesses (such as the Spicers, who claimed to have spied this mystery beast on land in 1933) have likened it to an enormous, hideous snail. Quite apart from the profound morphological modifications necessary for a beast corresponding to Shiels's elephant squid to have evolved from known cephalopod (squid and octopus) stock, however, a fundamental obstacle to this hypothetical creature's plausibility is that all known species of modern-day cephalopod are exclusively marine. There is not a single species of freshwater squid or octopus on record, and for one to evolve would require drastic tissue modifications relating to osmoregulatory ability.

Shiels's *Fortean Times* account attracted considerable interest within and beyond the fortean and cryptozoological fraternity, and summaries of his speculation subsequently appeared in a wide range of publications by other writers. Regrettably, however, many of these second-hand accounts mistakenly claimed that Shiels had formally dubbed his hypothetical elephant squid *Dinoteuthis proboscideus* (translating, incidentally, as 'trunked terrible squid'). In reality, conversely, as Shiels went on to explain in *Monstrum!*, Irish zoologist A.G. More had already given that particular name to a massive squid specimen beached at Dingle in County Kerry, Ireland, in October 1673 during a major storm. Instead, Shiels suggested that an apt name for his own, totally conjectural cephalopod would be *Elephanteuthis nnidnidi* – its species name needing no explanation for anyone knowing of Shiels's experiments with psychic automatism.

Pictish beast depicted on the east side of the Maiden Stone in a photograph (© Ronnie Leask/Wikipedia http://creativecommons.org/licenses/by-sa/2.0/) and a line drawing (public domain)

More recently, mystery beast researcher Scott Mardis from the USA has suggested that the Pictish beast images may actually depict an evolved, surviving species of short-necked plesiosaur (and therefore quite probably a pliosaur, which also had long jaws like those of the Pictish beast). Plesiosaurs have of course been officially extinct for at least 66 million years, but an evolved, surviving representative of the long-necked, short-jawed version (elasmosaur) of these aquatic prehistoric reptiles nevertheless has long been a popular cryptozoological identity for Nessie-type water monsters, as reviewed in Chapter 2 of this present book.

Alongside a depiction of a pliosaur, showing its long jaws and short neck
(© Dr Karl Shuker)

In short, the Pictish beast remains the subject of several interesting interpretations, but no satisfactory solutions - unless of course the answer lurks not among its petroglyphic portrayals but instead within the secretive depths of the lochs forming a major, familiar part of the landscape once inhabited by the painted people of Scotland's distant past?

Grace Connolly's tombstone, on which the dobhar-chú is portrayed
(© Dave Walsh)

CHAPTER 16:
THE IRISH MASTER OTTER IN SCOTLAND?

I would like to be very clear, before I go any further, that when I speak of an otter-like animal I am *not* saying it is an otter. Nor am I pretending to identify it in a scientific sense with any known group of animals, reptile, mammal or anything else. Nor am I saying that beyond a doubt there is such an animal. There are no grounds yet for anything more than speculation. All I am saying is that if there is such an animal its shape appears to approximate to that of an otter, and that its habits could be similar to those of an otter. It is only by making this kind of assumption and then speculating from it that we can hope to explain how a suspected animal believed to be 20 feet long could escape frequent detection.

Those who have made a study of otters in the wild know that they are probably the most elusive animal in the countryside. That, at least, is my experience. An otter may work a river near a village and nobody be aware of its presence...

Let us suppose that the habit and habitat of such a long-necked otter-like animal haunting Loch Ness agree with those of the common otter. Then we have to deal with a most elusive beast, hunting mainly inshore, perhaps basking at times at the water's edge, which for long stretches is out of sight except to the person who, very occasionally, takes the trouble to walk along it. Possibly it may go up the rivers and burns, but wherever it may go there are a thousand and one hiding places where even an animal of these proportions could lie hidden, or could move about without exposing itself unduly, especially if it were mainly nocturnal. If we argue that such an animal would be bound to be seen sooner or later, even in so sparsely populated an area - well, that is the kind of frequency with which it has been reported.

Maurice Burton – *The Elusive Monster*

In my book *The Beasts That Hide From Man* (2003), I devoted an extensive chapter to the master otter or dobhar-chú of Ireland, which was the very first comprehensive cryptozoological examination of this enigmatic mystery beast. I have also included coverage of it in other publications since then, including the following concise entry that I wrote for

Chambers Dictionary of the Unexplained (2007) in my capacity as this fact-filled tome's cryptozoological contributor:

Dobhar-chú
Savage otter-like lake monster from Ireland, depicted on the gravestone of an alleged victim.
Congbháil (Conwall) Cemetery in the town of Drumáin (Drummans), forming part of the approach to the Valley of Glenade in Ireland's County Sligo, contains a grave of considerable cryptozoological interest. The grave is that of Grace Connolly, who was allegedly killed one morning in September 1722 by a very large and savage otter-like lake monster known as the dobhar-chú ('water hound') or master otter. According to local lore, it had emerged from Glenade Lake (just inside County Leitrim's border with Sligo) while Grace was nearby, and fiercely attacked her. Her husband, Terence, later found her dead, bloodstained body at the lakeside, with the dobhar-chú lying across her. Enraged, he shot the beast dead, but before it died the creature let out a shrill scream and moments later a second dobhar-chú, presumably its mate, rose up from the lake's depths and chased after Terence, who fled on horseback. Finally, however, Terence succeeded in stabbing the vengeful dobhar-chú to death – and depicted on Grace's tombstone is the very act of the dobhar-chú being mortally stabbed in the chest by Terence, with its head thrown backwards in its death throes.

This tombstone portrays the dobhar-chú as being decidedly canine in overall form (thus explaining its Gaelic name's derivation), with long limbs, muscular haunches, deep chest, and a long tufted tail, but combined with these features are others that are undeniably otter-like, such as its tiny ears, very large paws, short head, and fairly long heavy neck – hence 'master otter'. Collectively, they yield an animal unlike anything known to contemporary zoology. Intriguingly, however, cryptozoology can offer a very comparable, recent version.

On 1 May 1968, John Cooney and Michael McNulty witnessed an extremely strange creature run across the road just in front of their vehicle and vanish into some undergrowth, as they were driving home past Sraheens Lough – a lake on Achill Island, off the western coast of County Mayo, which is in the same region of Ireland as Counties Leitrim and Sligo. They later described this animal as having four well-developed legs on which it rocked from side to side as it ran, a long sturdy tail, small head, and lengthy neck. Shiny dark-brown in colour, it measured 2.4-3.0 m in total length. During the next few weeks, several similar reports were made in this vicinity by others too. Could this cryptid have been a dobhar-chú?

There have been many claims of lake monsters inhabiting various of Ireland's loughs, but as many of these bodies of water are very small, sceptics have dismissed such a possibility by claiming that the loughs could not sustain such creatures. If, however, they have the ability to move from one lough to another, rather than residing permanently in any single body of water, lough size would not be a problem – and perhaps that is what the Sraheens Lough creature was doing when sighted. The nature of its zoological identity, meanwhile, and that of Glenade Lake's morphologically-reminiscent dobhar-chú too remain a complete mystery.

Moreover, such reports are not confined to Ireland, because sightings of a creature similar to Ireland's master otter have also emerged occasionally from mainland Scotland. Brought to my attention a while ago by correspondent Leslie Thomson, one such report is a noteworthy excerpt from *The History of the Scots From Their First Origin* by Hector Boece (English edition, 1575), and reads as follows:

> ...on the summer solstice of the year 1510 some kind of beast the size of a mastiff emerged at dawn from one of those lochs, named Gairloch, having feet like a goose, that without any difficulty knocked down great oak trees with the lashings of its tail. It quickly ran up to the huntsmen and laid low three of them with three blows, the remainder making their escape among the trees. Then, without any hesitation, it immediately returned into the loch. Men think that when this monster appears it portends great evil for the realm, for otherwise it is rarely seen.

Loch Gairloch is a sea loch on Scotland's northwest coast; it measures approximately 6 miles long by 1.5 miles wide. As for the creature that emerged from it, I think it safe to assume that its tail's oak-felling prowess owes more to literary exaggeration than to anatomical accuracy. Conversely, the likening of its feet to those of a goose probably indicated merely that they were webbed. Overall, therefore, the mastiff-sized, web-toed, fleet-footed, quadrupedal water monster of Gairloch does recall the master otter of Glenade Lake, but its taxonomic identity, as with the latter beast's, remains unresolved. Could the explanation simply be an extra-large version of the common otter *Lutra lutra*? Or are the master otter's lengthier limbs and other morphological differences evidence that it was – or is - an entirely separate, zoologically-undescribed species?

Similarly, as quoted at this present chapter's opening, British zoologist and Nessie researcher Dr Maurice Burton cautiously speculated that the LNM may also be a novel species of long-necked otter. He even devoted a entire chapter to this possibility in his book *The Elusive Monster* (1961).

Leslie also sent me a second, equally remarkable excerpt from Boece's book:

> In this estuary there sometimes appear some ill-omened phantoms with human faces, wearing monks' cowls, as it seems, which rise out of the water as far as their waists, and these are called bassinats in our native language.

Rising up out of the water as far as their waists is an activity that readily calls to mind inquisitive seals, but seals do not wear cowls or possess human faces, so what could the bassinats be? I have been unable to locate any definition or information relating to this term – but in a response of 23 February 2011 to this chapter when originally posted on my *ShukerNature* blog as an article on 20 February 2011, reader Chris Woodyard offered the following insight and suggestions:

> A monk's cowl, as it was understood in the 16th century, did not just mean a hood over the head. The cowl, in monastic terms, is a hood and a short shoulder cape. When you see a monk with a short cape and his hood thrown back, bunched around his neck, that is still a cowl, which suggests to me that the author was describing rolls of fat/blubber/skin around the creature's neck. As for the human face, I have no idea, but there are all those dugong/manatee mermaid stories and seals can look adorably human at times. A "Monk" seal?

Interestingly, there are indeed such creatures as monk seals, a trio of species belonging to the genus *Monachus*. They derive their name from the noticeable amount of skin and fat around their head and neck, which resembles a monk's cowl – exactly as Chris speculated in relation to the mysterious bassinats.

Moreover, one of these three species – the Mediterranean monk seal *M. monachus* - is European in distribution (the other two are Hawaiian and Caribbean respectively, though the Caribbean species became extinct during the 20th Century), but is now critically endangered. According to its entry in the IUCN Red List's website, however, this species was once:

> ...widely and continuously distributed in the Mediterranean, Black and adjacent seas, and in the North Atlantic waters from Morocco to Cap Blanc, including the Canary Islands, Madeira Islands and the Azores. A few individuals, possibly vagrants, have been recorded in Senegal, the Gambia and the Cape Verde Islands in the south end of the distribution range, as well as in Portugal and Atlantic France in the northern end.

Could it be, therefore, that over 500 years ago, when this species was still common and widely distributed, a few vagrants made it as far north as the coasts of Scotland, explaining the cowled bassinats? Returning to the term 'bassinat' – is it familiar to anyone reading this chapter? If so, I'd greatly welcome details – thanks!

As for the most famous Scottish water cryptid of all, Nessie the LNM, its multi-faceted history draws now to a close in this present book, but somewhere out there, amid a vast expanse and volume of dark, cold, peaty water, it continues - a saga of centuries, an ongoing mystery captivating each new generation that learns of it, just as it captivated me when I first read about this cryptozoological mega-star in my *Stranger Than People* book that my dear mother bought me all those many years ago when I was a child.

Will it ever be solved? Is there truly a Nessie species awaiting zoological discovery and description?

I cannot answer those questions, but who knows – perhaps someone reading this present book of mine right now might be sufficiently inspired by its contents to go

Monk seal vertical in water (public domain)

forth and not only seek but also finally uncover the long-awaited solution to this (very) longstanding riddle. And should that ever happen, I shall finally be reassured that my years in this world have indeed achieved something, which is all that anyone can ever hope for.

Having said that: as is true of so much else in life, such an outcome may well result not from choice but simply from chance, not from labour but merely from luck. As veteran LNM researcher Prof. Henry H. Bauer so adroitly commented in his own book *The Enigma of Loch Ness* (1986):

> In Nessiedom, in part because it is exploration as much as experimentation, luck and chance play a huge and possibly even decisive role. Indeed, it is entirely possible that the definitive proof of Nessie's existence will come from some naïve tourist, entirely ignorant of the history of the quest, perhaps even equipped with ridiculously expensive gadgetry which he doesn't know how to use properly but which nevertheless does the job.

Ah well, t'was ever thus!

From Samuel Wendell Williston's book *Water Reptiles of the Past and Present* (1914), an exquisite early 20th-Century restoration of Mesozoic aquatic life, showcasing a very elegant, graceful, swan-necked plesiosaur – thereby exemplifying a classic zoological image that has transcended palaeontological exactitudes to become an art icon in its own right, and also, rightly or wrongly, the visual embodiment of Nessie (public domain)

A BIBLIOGRAPHY OF NESSIE-THEMED
NON-FICTION BOOKS

In 1934, just over 80 years ago now, a humble, unassuming little pamphlet entitled *The Home of the Loch Ness Monster*, a mere 18 pages long and containing only a single illustration, was privately published by its author, a retired British army man named Lieutenant-Colonel William Horsburgh Lane who lived in a house on the shores of Loch Ness. It documented Lane's considered belief that the LNM may be a giant form of salamander. This particular zoological identity would be revisited and revitalised more than four decades later by Chicago University biochemist and veteran Nessie researcher Prof. Roy P. Mackal in his own much more detailed, highly-acclaimed study, *The Monsters of Loch Ness* (1976), which remains a standard work on the subject.

Conversely, Lane's very modest contribution to the LNM literature is all but forgotten today and exceedingly rare, making it much sought-after by Nessie aficionados – but not only because if its scarcity. What gives his pamphlet a unique, unassailable, and truly historic place within the cryptozoological canon, and the reason why I have highlighted it here, is that it was the very first entire, stand-alone work (i.e. book, pamphlet, magazine, as opposed to merely an article in a magazine or a report in a newspaper) devoted exclusively to the LNM.

This, therefore, is where it all began – the trickle that became a stream that became a river that became a veritable ocean of Nessie-themed books published in the eight decades that have followed in the wake of Lane's pamphlet. And what a number there have been, as I shall now reveal here in this comprehensive LNM bibliography.

NB – There are countless non-fiction books that include a chapter or section concerned with the LNM, so (with one notable exception) this present bibliography confines itself to those non-fiction books that are entirely devoted to Nessie, or to other Scottish loch monsters. The one notable exception is Peter Costello's lake monsters book, which does contain a few chapters towards the end concerning other aquatic cryptids but is predominantly Nessie-based in content and like Mackal's book is widely recognised as a standard work on the LNM.

- ANON., *They Saw Nessie (Or Thought They Did!): Eye-Witness Sightings of the Monster Over the Years* (Northern Books, from Famedram: Gartocharn, 1984).
- ABELS, Harriette & SCHROEDER, Howard, *The Loch Ness Monster* (Crestwood House: New York, 1987).
- AKINS, William, *The Loch Ness Monster* (Signet: New York, 1977).
- ARMSTRONG, Edward, *Sticking My Neck Out* (Privately published: [no place of publication details], 1983).
- BAUER, Henry H., *The Enigma of Loch Ness: Making Sense of a Mystery* (University of Illinois Press: Urbana, 1986).
- BAUMANN, Elwood D., *The Loch Ness Monster* (Franklin Watts: New York, 1972).
- BAXTER, Colin, *The Loch Ness Monster* (Colin Baxter Photography: Grantown on Spey, 2012).
- BENDICK, Jeanne, *The Mystery of the Loch Ness Monster* (McGraw-Hill: London, 1976).
- BERTON, Jean, *Les Monstres du Loch Ness et d'Ailleurs* (France-Empire: Paris, 1977).
- BINNS, Ronald, *The Loch Ness Mystery Solved* (Open Books: Shepton Mallet, 1983).
- BORDER, Rosemary, *Loch Ness Monster* (Macdonald Phoebus: London, 1979).
- BRASSEY, Richard, *Nessie the Loch Ness Monster* (Orion: London, 2010).
- BURTON, Maurice, *The Elusive Monster: An Analysis of the Evidence From Loch Ness* (Rupert Hart-Davis: London, 1961).
- BYRNE, Gerald, *Gestalt Forms of Loch Ness* (JRP Ringler: Zurich, 2011).
- CAMPBELL, Elizabeth M. & SOLOMON, David, *The Search For Morag* (Tom Stacey: London, 1972).
- CAMPBELL, Steuart, *The Loch Ness Monster: The Evidence* (Aquarian: Wellingborough, 1986; rev. edit., Aberdeen University Press: London, 1991).
- CARNEY, James, *Loch Ness Monster* (Colin Baxter Photography: Grantown on Spey, 2009).
- CARRUTH, J.A., *Loch Ness and Its Monster* (Abbey Press: Fort Augustus, 1945).
- CASSIE, R.L., *The Monsters of Achanalt* (2 vols) (D. Wyllie & Sons: Aberdeen, 1935-36).
- COOK, D. & COOK, Y., *The Great Monster Hunt: The Story of the Loch Ness Investigation* (Grosset & Dunlap: New York, 1969).
- CORNELL, James, *The Monster of Loch Ness* (Scholastic Book Services: New York, 1977).
- COSTELLO, Peter, *In Search of Lake Monsters* (Garnstone Press: London, 1974).
- CRISOLOGO, Jonalyn, *et al.*, *The Loch Ness Monster For Kids: Fact or Fiction?* (CreateSpace: Scotts Valley, 2014).
- DINSDALE, Angus, *The Man Who Filmed Nessie* (Hancock House: New York, 2013).
- DINSDALE, Tim, *Loch Ness Monster* (Routledge & Kegan Paul: London, 1961; 4th edit., 1982).
- DINSDALE, Tim, *The Story of the Loch Ness Monster* (Allan Wingate: London, 1973).
- DINSDALE, Tim, *Project Water Horse: The True Story of the Monster Quest at Loch Ness* (Routledge & Kegan Paul: London, 1975).
- FLEMING, Maureen & KING, Virginia, *The Loch Ness Monster Mystery* (Mimosa Publications: Hawthorn, 1995).

- FLITCROFT, Jean, *The Loch Ness Monster* [The Cryptid Files #1] (Darby Creek Publishing: Minneapolis, 2014).
- GALLAGHER, Patrick J., *Loch Ness: From Out of the Depths – Original Newspaper Accounts of the Rise of the Loch Ness Monster 1933-1934* (CreateSpace: Scotts Valley, 2015).
- GANTES, Rémy, *Le Mystère du Loch Ness* (Etudes Vivantes: Paris, 1979).
- GIBSON, John A. & HEPPELL, David (Eds), *Proceedings of the Symposium on the Loch Ness Monster: "The Search For Nessie in the 1980s"* (Scottish Natural History Library: Foremount House, 1988).
- GOULD, Rupert T., *The Loch Ness Monster and Others* (Geoffrey Bles: London, 1934).
- GRIMSHAW, Roger & LESTER, Paul, *The Meaning of the Loch Ness Monster* (Birmingham University: Birmingham, 1976).
- HAMILTON, W.D. & HUGHES, J., *The Mysterious Monster of Loch Ness* (Fort Augustus Abbey Press: Fort Augustus, 1934).
- HANSEN, Kim M., *Mysteriet om Nessie: Søslangen i Loch Ness* (Gyldendal: Copenhagen, 1988).
- HARMSWORTH, Tony, *The Mysterious Monsters of Loch Ness* (Precision Press: 1980).
- HARMSWORTH, Tony, *Loch Ness: The Monster* (Peter Gray Ltd: Tillicoultry, 1985).
- HARMSWORTH, Tony, *Loch Ness, Nessie & Me: The Truth Revealed* [vt *Nessie Understood*] (Harmsworth.net: Drumnadrochit, 2010).
- HARMSWORTH, Tony, *Loch Ness Understood* (Harmsworth.net: Drumnadrochit, 2012).
- HARRISON, Paul, *The Encyclopedia of the Loch Ness Monster* (Robert Hale: London, 1999).
- HASTAIN, Ronald & WITCHELL, Nicholas, *Loch Ness and the Monster: A Handbook For Tourists* (J. Arthur Dixon: Inverness, 1971).
- HAUF, Monika, *Nessie – Das Ungeheur von Loch Ness* (Bohmeier Verlag: Leipzig, 2003).
- HILE, Lori, *The Loch Ness Monster* (Capstone Global Library: Oxford, 2013).
- HOLIDAY, F.W., *The Great Orm of Loch Ness: A Practical Inquiry Into the Nature and Habits of Water-Monsters* (Faber & Faber: London, 1968).
- JAMES, David, *Loch Ness Investigation* (Loch Ness Phenomena Investigation Bureau: London, 1968).
- KALLEN, Stuart A., *The Loch Ness Monster* (Referencepoint Press: San Diego, 2008).
- KIRKPATRICK, Betty, *Nessie: The Legend of the Loch Ness Monster* (Crombie Jardine Publishing: Cheam, 2005).
- KLEIN, Martin, *et al.*, *Underwater Search at Loch Ness* [Monograph No. 1] (Academy of Applied Science: Belmont, 1972).
- LANE, W[illiam].H., *The Home of the Loch Ness Monster* (Moray Press: Edinburgh, 1934).
- MacRAE, Jim, *Loch Ness Monster Handbook* (John G. Eccles: Inverness, 1974).
- MACKAL, Roy P., *The Monsters of Loch Ness* (Macdonald and Janes: London, 1976).
- MARTIN, David & BOYD, Alistair, *Nessie: The Surgeon's Photograph Exposed* (Martin and Boyd: East Burnet, 1999).
- MEREDITH, Dennis L., *Search at Loch Ness: The Expedition of the New York Times*

and the Academy of Applied Science (Quadrangle: New York, 1977).

* MILLER, Connie C., *El Monstruo del Lago Ness: El Misterio Sin Resolver / The Loch Ness Monster: The Unsolved Mystery* (Blazers Bilingual: North Mankato, 2009).
* MUNRO, Donald J., *Loch Ness Mystery* (Privately published: [no place of publication details], 1937).
* OUDEMANS, Anthonie C., *The Loch Ness Animal* (E.J. Brill: Leiden, 1934).
* OWEN, William, *Loch Ness Revealing Its Monsters* (Jarrold: Norwich, 1976).
* OWEN, William, *The Loch Ness Monster* [vt *Scotland's Loch Ness Monster*] (Jarrold: Norwich, 1986).
* PARKS, Peggy J., *The Loch Ness Monster* (KidHaven: Farmington Hills, 2006).
* PERERA, Victor, *The Loch Ness Monster Watchers* (Capra Press: Santa Barbara, 1974).
* PICKNETT, Lynn, *The Loch Ness Monster* (Pitkin Pictorials: Andover, 1993).
* REDFERN, Nick, *Nessie: Exploring the Supernatural Origins of the Loch Ness Monster* (Llewellyn: St Paul, 2016).
* RICHARDS, Amber, *Legend of Loch Ness Monster For Kids: A Mystery in the United Kingdom* (CreateSpace: Scotts Valley, 2014).
* RUSSELL, Jessie & COHN, Ronald (Eds), *Loch Ness Monster in Popular Culture* [a hard-copy compilation of Wikipedia articles] (Bookvika Publishing: Moscow, 2012).
* RUSSELL, Jessie & COHN, Ronald (Eds), *Loch Ness Monster* [a hard-copy compilation of Wikipedia articles] (Bookvika Publishing: Moscow, 2012).
* SAN SOUCI, Robert D. (Ed.), *The Loch Ness Monster: Opposing Viewpoints* (Greenhaven Press: San Diego, 1989).
* SCHACH, David, *The Loch Ness Monster* (Torque Books: Minneapolis, 2010).
* SEARLE, Frank, *Nessie: Seven Years in Search of the Monster* (Coronet: London, 1976).
* SEARLE, Frank, *The Story of Loch Ness* (John Eccles: Inverness, 1977).
* SEARLE, Frank, *Around Loch Ness: A Handbook For Nessie Hunters* (John Eccles: Inverness, 1977).
* SENSICAL, Benjamin, *Loch Ness: An Explanation* (Privately published: [no place of publication details], 1982).
* SHINE, Adrian, *Loch Ness?* (Loch Ness Project: Drumnadrochit, 2006).
* SHUKER, Karl P.N., *Here's Nessie! A Monstrous Compendium From Loch Ness* (CFZ Press: Bideford, 2016).
* SIEVERT, Terri, *The Unsolved Mystery of the Loch Ness Monster* (First Fact Books: North Mankato, 2013).
* SMITH, Warren, *Strange Secrets of the Loch Ness Monster* (Kensington Publishing Corporation: New York, 1976).
* SNYDER, Gerald S., *Is There a Loch Ness Monster? The Search For a Legend* (Julian Messner: New York, 1977).
* THORNE, Ian, *The Loch Ness Monster* (Crestwood House: New York, 1978).
* TROUPE, Thomas K., *The Legend of the Loch Ness Monster* (Picture Window Books: North Mankato, 2011).

- VIBE, Palle, *Gaden I Loch Ness* (Rhodos: Copenhagen, 1970).
- WALLACE, Holly, *The Mystery of the Loch Ness Monster* (Capstone Global Library: Oxford, 2006).
- WATSON, Roland, *The Water Horses of Loch Ness* (CreateSpace: London, 2011).
- WELSH, Lily (Ed.), *Monsters and Myths: Loch Ness and the Monster* [a hard-copy compilation of Wikipedia articles] (Webster's Digital Services: [no place of publication details], 2010).
- WHYTE, Constance, *The Loch Ness Monster* (Headley Brothers: London, 1951).
- WHYTE, Constance, *More Than a Legend: The Story of the Loch Ness Monster* (Hamish Hamilton: London, 1957; rev. 1961).
- WILLIAMS, Gareth, *A Monstrous Commotion: The Mysteries of Loch Ness* (Orion Books: London, 2015).
- WITCHELL, Nicholas, *The Loch Ness Story* (Terence Dalton: Lavenham, 1974; 3rd edit., Corgi Books: London, 1989).
- WITCHELL, Nicholas, *Loch Ness and the Monster* (J. Arthur Dixon: Newport, 1975).

ADDITIⱭNAL BIBLIⱭGRAPHY
FⱭR THIS BⱭⱭK

During this book's preparation, I was able to consult the vast majority of those titles already included in the earlier bibliography of books devoted to the LNM (and which are therefore not included again below). However, the following publications were also of great use to me, so I am listing them here in this additional bibliography.

- ANON., 'The Sea Serpent in the Highlands', *Inverness Courier* (Inverness), 6 March (1856).
- ANON., 'A Diver's Experience – A Strange Creature – Strange Creatures', *Northern Chronicle* (Inverness), 31 January (1934).
- ANON., 'Camera Catches the Loch Ness Monster – Views of Monster Lessen Skepticism' [re Surgeon's Photo]', *New York Times* (New York), 22 April (1934).
- ANON., 'Police Seize ½-Ton Creature 'Found at Loch Ness", *Daily Telegraph* (London), 1 April (1972).
- ANON., 'Loch Ness Yields Another One – 12ft Long, With Fur', *Times* (London), 1 April (1972).
- ANON., "Nessie' - It's a Frozen Load of Bull', *Wolverhampton Express and Star* (Wolverhampton), 1 April (1972).
- ANON., 'I Put 'Nessie' in the Loch!', *News of the World* (London), 2 April (1972).
- ANON., 'Hoax Seal On View', *Times* (London), 3 April (1972).
- ANON., 'Contact! Nessie Hunters Find 'Something Big' 200ft Down', *Evening Times* (London), 9 October (1987).
- ANON., 'Jeg var Søslangen i Loch Ness', *Hjemmet*, No. 47 (1992).
- ANON., 'Cavern May Be Lair of Nessie', *Daily Post* (Liverpool), 19 February (1997).
- ANON., 'Experts Say Nessie Fossil is a Hoax', *Inverness Courier* (Inverness), 18 July (2003).
- ANON., 'Has the Loch Ness Monster's Tooth Been Found?', *Loch Ness Tooth*, http://www.lochnesstooth.com/ (2011).
- ANON., 'Nessie's Underwater Hideaway? Video Shows World Famous Loch Ness Has

Just Got Even Deeper', *Daily Record* (Glasgow), http://www.dailyrecord.co.uk/news/scottish-news/nessies-underwater-hideaway-video-shows-7203518#ZaVmCrThos7mB67d.97 19 January (2016).

- ANON., 'A Real Monster', *Loch Ness & Morar Project*, http://www.lochnessproject.org/explore_loch_ness/MUNIN%20SURVEY/Loch%20Ness%20Project%20Munin.html April (2016).

- ANSTRUTHER-GRAY, W., 'More About the Monster [re Loch Arkaig monster]', *Times* (London), 14 December (1933).

- BARKER, Will, 'Nessie is a 100-Yr-Old Eunuch Eel', *Sun* (London), 22 September (2003).

- BAUER, Henry H., 'The Case For the Loch Ness "Monster": The Scientific Evidence', *Journal of Scientific Exploration*, 16(2): 225-246 (2002); 'Genuine Facts About "Nessie", the Loch Ness "Monster"', *Henry H. Bauer – Homestead*, http://henryhbauer.homestead.com/LochNessFacts.html 30 November (2014).

- BERNARD, Aurélien, *et al.*, 'Regulation of Body Temperature by Some Mesozoic Marine Reptiles', *Science*, 328 (11 June): 1379-1382 (2010).

- BOECE, Hector, *The History of the Scots From Their First Origin* (English edit.) (London, 1575).

- BOND, Anthony, 'Loch Ness Monster 'Lair' Does NOT Exist After State-of-the-Art Robot Searches Depth of Famous Lake', *Mirror* (London), http://www.mirror.co.uk/news/weird-news/loch-ness-monster-lair-not-7743201 13 April (2016).

- BORD, Janet & BORD, Colin, *Ancient Mysteries of Britain* (Grafton Books: London, 1989).

- BORROW, George, *Wild Wales: Its People, Language, and Scenery* (3 vols) (John Murray: London, 1862).

- BRAITHWAITE, Hugh, 'Sonar Picks Up Stirrings in Loch Ness', *New Scientist*, 40 (19 December): 664-666 (1968).

- BRIGGS, Katharine, *A Dictionary of Fairies* (Allen Lane: London, 1976).

- BURTON, Maurice, 'Loch Ness Monster: A Reappraisal', *New Scientist*, 8 (22 September): 773-775 (1960).

- CAMPBELL, J.F., *Popular Tales of the West Highlands* (4 vols) (Edmonston & Douglas: Edinburgh, 1860-1862).

- CAMPBELL, John G., *Superstitions of the Highlands and Islands of Scotland* (James MacLehose: Glasgow, 1900).

- CHAMPAGNE, Bruce A., 'A Classification System For Large, Unidentified Marine Animals Based on the Examinations of Reported Observations', *In:* HEINSELMAN, Craig (Ed.), *Elementum Bestia* (Lulu: Raleigh, 2007): 144-172.

- CHORVINSKY, Mark, 'Nessie the Loch Ness Monster', *Fate*, 46 (November): 48-55 (1993).

- COLEMAN, Loren & HUYGHE, Patrick, *The Field Guide to Lake Monsters, Sea Serpents, and Other Mystery Denizens of the Deep* (Tarcher/Penguin: New York, 2003).

- COONEY, Jay, 'The Mysterious "White Mice" of Loch Ness: Larval Longnecks?', *Bizarre Zoology*, http://bizarrezoology.blogspot.com/2014/03/the-mysterious-mice-of-loch-ness-larval.html 30 March (2014) [no longer online].

- CORNES, Robert, 'The Seal Serpent: The Case For The Surreal Seal', *In:* DOWNES, Jonathan (Ed.), *CFZ 2007 Yearbook* (CFZ Press: Bideford, 2007): 83-199.
- CRUICKSHANK, Arthur, *et al.*, 'Dorsal Nostrils and Hydrodynamically Driven Underwater Olfaction in Plesiosaurs', *Nature*, 352 (4 July): 62-64 (1991).
- DASH, Mike, 'Cryptozoology Conference 1987', *Fortean Times*, 49 (winter): 36-37 (1987); 'Footnote to a Footnote [re monsters of Achanalt]', *Fortean Times*, No. 52 (summer): 66-67 (1989); 'Loch Ness Eels', *Fortean Times*, No. 212 (July): 74-76 (2006); 'Monsters of Achanalt', *Fortean Times*, No. 177 (winter): 48 (2007).
- DINSDALE, Tim, *The Leviathans* (Routledge & Kegan Paul: London, 1966; rev. edit., Futura: London, 1976); 'The Rines/Edgerton Picture', *Photographic Journal*, (April): 162-165 (1973).
- DOWNES, Jonathan, 'Not On Your Nessie! [re eunuch eel identity theory]', *Fortean Times*, No. 182 (April): 58-59 (2004).
- EBERHART, George M., *Mysterious Creatures: A Guide to Cryptozoology* (2 vols) (ABC-Clio: Santa Barbara, 2002; 2nd edit. (2 vols), CFZ Press: Bideford, 2013, 2015).
- ECKERT, S.A., 'Swim Speed and Movement Patterns of Gravid Leatherback Sea Turtles (*Dermochelys coriacea*) at St Croix, US Virgin Islands', *Journal of Experimental Biology*, 205: 3689–3697 (2002).
- EVANS, Mark, 'An Investigation Into the Neck Flexibility of Plesiosauroid Plesiosaurs: *Cryptoclidus eurymerus* and *Muraenosaurus leedsii*', M.Sc. Research Project in Vertebrate Palaeontology, University College London, May 1993 (minor corrections September 2014).
- FORDYCE, L. McP., 'Was it the Monster? [re 'camel-horse' sighting]', *Scots Magazine (New Series)*, 132 (June): 269-277 (1990).
- FRAIR, W., *et al*,. 'Body Temperature of *Dermochelys coriacea*: Warm Turtle From Cold Water', *Science*, 177 (1 September): 791-793 (1972).
- FRASER, Iain, *The Pictish Symbol Stones of Scotland* (Royal Commission on the Ancient and Historic Monuments of Scotland: Edinburgh, 2008).
- GOULD, Rupert T., *The Case For the Sea Serpent* (Philip Allan: London, 1930).
- GREEN, Roger L., *Myths From Many Lands* (Purnell: London, 1965).
- GREENWELL, J. Richard, 'An Endothermic "Nessie"?', *Science*, 200 (19 May): 722-723 (1978).
- GREENWELL, J. Richard (Ed.), 'Retouching of Nessie Flipper Photo Claimed – Denied', *ISC Newsletter*, 3 (No. 4; winter): 1-5 (1984); 'Hoax Claimed For Wilson Photo', *ISC Newsletter*, 3 (No. 4; winter): 5-6 (1984); 'Nessie Symposium Highlight of Edinburgh Meeting', *ISC Newsletter*, 6 (No. 4; winter): 1-5 (1987); 'Nessie Survives Deepscan Sonar Probe', *ISC Newsletter*, 7 (No. 1; spring): 5-7 (1988).
- GREW, Nehemiah, *Musaeum Regalis Societatis: Or a Catalogue and Description of the Natural and Artificial Rarities Belonging to the Royal Society and Preserved at Gresham Colledge* [sic] (W. Rawlins: London, 1681).
- HAINING, Peter, *The Monster Trap and Other True Mysteries* (Armada: London, 1976).
- HANLON, Joseph, 'The Pictures Show Something, But Is It Nessie?', *New Scientist*, 68

(18-25 December): 738-739 (1975).

- HEPPLE, Rip (Ed.), *Nessletter*, Nos 1-current issue (1974-present day).
- HEUVELMANS, Bernard, *Le Grand Serpent-de-Mer* (Plon: Paris, 1965); *In the Wake of the Sea-Serpents* (Rupert Hart-Davis: London, 1968).
- HISTORIC ENVIRONMENT SCOTLAND, *Pictish Stones*, http://www.pictishstones.org.uk/pictishstones/pictishstoneshome.htm (n.d. – most recently accessed by me on 4 May 2016).
- HOLIDAY, F.W. & WILSON, Colin, *Goblin Universe* (Llewellyn Publishers: St Paul, 1986).
- KELLOCK, Andrew, 'More Sightings of Loch Ness Monster in 2015 Than Anytime in Last Ten Years', *Press & Journal* (Aberdeen), https://www.pressandjournal.co.uk/fp/news/inverness/773587/more-sightings-of-loch-ness-monster-in-2015-than-anytime-in-last-ten-years/ 8 December (2015).
- LANGELAAN, George, *Les Faits Maudits* (Encyclopédie Planète: Paris, 1967).
- LANGTON, James, 'Revealed: The Loch Ness Picture Hoax', *Sunday Telegraph* (London), 13 March (1994).
- LAUDER, James, 'The Minor Monsters', *In Britain*, 28 (April): 17-18 (1973).
- LeBLOND, Paul H. & COLLINS, Michael J., 'The Wilson Nessie Photo: A Size Determination Based On Physical Principles', *Cryptozoology*, 6: 55-64 (1987).
- LIU, S., *et al.*, 'Computer Simulations Imply Forelimb-Dominated Underwater Flight in Plesisaurs', *PLoS Computational Biology*, 11 (18 December): 1-18 (2015).
- LUCAS, Spencer B. & REYNOLDS, Robert E., 'Putative Plesiosaurs From Cajon Pass, California, U.S.A.', *Cretaceous Research*, 14: 107-111 (1993).
- McCORMICK, Cameron, 'The Unknown Sea Serpent Classification', *The Lord Geekington*, http://cameronmccormick.blogspot.co.uk/2008/01/unknown-sea-serpent-classification.html 24 January (2008); 'The Flexibility of Plesiosaur Necks', *The Lord Geekington*, http://cameronmccormick.blogspot.co.uk/2008/09/flexibility-of-plesiosaur-necks.html 28 September (2008); 'Weddell's Long-Necked Seal', *Biological Marginalia*, https://biologicalmarginalia.wordpress.com/2013/03/04/weddell-long-necked-seal/ 4 March (2013); 'The Hidden Necks of Seals', *Biological Marginalia*, https://biologicalmarginalia.wordpress.com/2013/03/15/the-hidden-necks-of-seals/ 15 March (2013).
- McCOY, Victoria E., *et al.*, 'The 'Tully Monster' is a Vertebrate', *Nature*, 532 (28 April): 496-499 (2016).
- McEWAN, Graham J., *Mystery Animals of Britain and Ireland* (Robert Hale: London, 1986).
- McGOVERN, Una (Ed.), *Chambers Dictionary of the Unexplained* (Chambers: Edinburgh, 2007).
- McKENZIE, Steven, 'Pictish Beast Intrigues Highland Archaeologists', *BBC News*, http://www.bbc.co.uk/news/uk-scotland-highlands-islands-14881753 14 September (2011); 'Film's Lost Nessie Monster Prop Found in Loch Ness', *BBC News*, http://www.bbc.co.uk/news/uk-scotland-highlands-islands-36024638 13 April (2016).
- MACKAL, Roy P., *Searching For Hidden Animals* (Doubleday: Garden City, 1980).
- MACKLIN, John, 'The Trap He Set Was For a Monster...But It Was the Colonel Who

Died', *Leicester Mercury* (Leicester), 28 March (1966).

- MAGIN, Ulrich, 'Captures and Strandings of Monsters in Loch Ness', *Pursuit*, 17 (Third Quarter): 121-122 (1984); 'St George Without a Dragon: Bernard Heuvelmans and the Sea Serpent', *Fortean Studies*, 3: 223-236 (1996); 'Waves Without Wind and a Floating Island: Historical Accounts of the Loch Ness Monster', *Fortean Studies*, 7: 95-115 (1999); 'Wattie', *Fortean Times*, No. 255 (November): 69 (2009).
- 'MANDRAKE' [PURSER, Philip], 'Making of a Monster', *Sunday Telegraph* (London), 7 December (1975).
- MANNING-SAUNDERS, Rosemary, *A Book of Monsters* (Piccolo Books: London, 1975).
- MARCELLA, 'Saint Mochua of Balla', *Omnium Sanctorum Hiberniae*, http:// omniumsanctorumhiberniae.blogspot.co.uk/2014/03/saint-mochua-of-balla-march-30.html 30 March (2014).
- MARDIS, Scott, 'Sealing Champ's Fate: More Thoughts on the Lake Monster', *Vox* (Vermont), 7 August (1996); 'Plesiosaurs on Ice: Perspectives on the "Living Plesiosaur" Controversy', *Bizarre Zoology*, http:// bizarrezoology.blogspot.co.uk/2014/05/plesiosaurs-on-ice-perspectives-on.html 9 May (2014); 'The Prehistoric Survivor Paradigm and More on Purported Post-Cretaceous Plesiosaurs', *Bizarre Zoology*, http://bizarrezoology.blogspot.co.uk/2014/07/the-prehistoric-survivor-paradigm-and.html 15 July (2014).
- MARSHALL, Michael, 'Lost Treasures: The Loch Ness Monster That Got Away', *New Scientist*, https://www.newscientist.com/article/mg21328502.300-lost-treasures-the-loch-ness-monster-that-got-away/?DCMP=OTC-rss&nsref=online-news 4 February (2012).
- MEURGER, Michel & GAGNON, Claude, *Lake Monster Traditions: A Cross-Cultural Analysis* (Fortean Tomes: London, 1988).
- MITCHELL, Victoria, 'Nessie Beats Burns To Be Crowned Most Famous Scot', *Scotsman* (Edinburgh), 29 November (2006).
- MOORE, Victoria, 'Was Nessie Really A Nelly? [re swimming elephant theory]', *Daily Mail* (London), 7 March (2006).
- MUIZON, Christian de, 'Les Vertébrés Fossiles de la Formation Pisco (Pérou). Première Partie: Deux Nouveaux Monachinae (Phocidae, Mammalia) du Pliocene de Sud-Sacaco', *Travaux de l'Insitut Français d'Études Andines*, 22: 1-161 (1981).
- NAISH, Darren, 'Sea Serpents, Seals, and Coelacanths', *Fortean Studies*, 7: 75-94 (2001); 'Swan-Necked Seals', *Tetrapod Zoology*, http:// darrennaish.blogspot.co.uk/2006/02/swan-necked-seals.html 4 February (2006); 'Plesiosaurs and the Repeated Invasion of Freshwater Habitats: Late-Surviving Relicts or Evolutionary Novelties?', *Tetrapod Zoology*, http://blogs.scientificamerican.com/ tetrapod-zoology/plesiosaurs-invaded-freshwater-habitats/ 9 January (2013); 'Photos of the Loch Ness Monster, Revisited', *Tetrapod Zoology*, http:// blogs.scientificamerican.com/tetrapod-zoology/photos-of-the-loch-ness-monster-revisited/ 10 July (2013); 'Plesiosaur Peril – The Lifestyles and Behaviours of Ancient Marine Reptiles', *Tetrapod Zoology*, http://blogs.scientificamerican.com/tetrapod-zoology/plesiosaur-peril-the-lifestyles-and-behaviours-of-ancient-marine-reptiles/ 3

March (2014).

- NEWTON, Michael, *Encyclopedia of Cryptozoology: A Global Guide* (McFarland: Jefferson, 2005).
- NICHOLLS, Elizabeth & RUSSELL, Anthony, 'The Plesiosaur Pectoral Girdle: The Case For a Sternum', *Neues Jahrbuch für Geologie und Paläontologie Abhandlungen*, 182: 161-185 (1991).
- O'KEEFE, F. Robin & CHIAPPE, Luis M., 'Viviparity and K-Selected Life History in a Mesozoic Marine Plesiosaur (Reptilia, Sauropterygia)', *Science*, 333: 870-873 (2011).
- OUDEMANS, A[nthonie].C., *The Great Sea-Serpent: An Historical and Critical Treatise* (Luzak & Co: London/E.J. Brill: Leiden, 1892).
- OWEN, James, 'Loch Ness Sea Monster Fossil a Hoax', *National Geographic News*, http://news.nationalgeographic.com/news/2003/07/0729_030729_lochness.html 29 July (2003).
- PARSONS, James, 'A Dissertaion Upon the Class of the Phocae Marinae, by James Parsons M.D.F.R.S.', *Philosophical Transactions*, 47 (1 January): 109-122 (1751).
- PAUL, William, 'Baffled Scientists Fail to Unravel Mystery of Nessie', *Scotsman* (Edinburgh), 27 July (1987).
- PAXTON, Charles G.M. & SHINE, Adrian J., 'A Checklist of Historical Hypotheses for the Loch Ness Monster', *Journal of Cryptozoology*, 4 (in press) (2016).
- POWELL, Chris, 'Killed By A Monster?', *Evening Sentinel* (Stoke-on-Trent), 2 September (1983).
- POWER, Dennis & JOHNSON, Donald, 'A Fresh Look at Nessie [re swimming elephant theory]', *New Scientist*, 83 (2 August): 358-359 (1979).
- RADFORD, Benjamin, 'Loch Ness Trench Spurs Monster Speculation', *Discovery News*, http://news.discovery.com/animals/loch-ness-trench-spurs-monster-speculation-160122.htm 22 January (2016).
- RADFORD, Benjamin & NICKELL, Joe, *Lake Monster Mysteries: Investigating the World's Most Elusive Creatures* (University Press of Kentucky: Lexington, 2006).
- RAYNOR, Dick, 'The Flipper Pictures Re-examined', *Loch Ness Investigation*, http://www.lochnessinvestigation.com/flipper.html 6 August (2002; rev. May 2009).
- REYNOLDSON, T., *et al.*, 'A Species of North American Triclad (Paludicola; Turbellaria) New to Britain Found in Loch Ness, Scotland', *Journal of Zoology*, 193 (April): 531-539 (1981).
- RICKARD, Robert J.M., 'Swedish Mystery Subs', *Fortean Times*, No. 42 (autumn): 7-8 (1984).
- ROBINSON, Anne & DAWE, Tony, "I Dumped Nessie in the Loch", *Sunday Times* (London), 2 April (1972).
- ROSIE, George, 'Something New Stirs at Bottom of Loch Ness', *Sunday Times* (London), 20 September (1981).
- ROSS, John, 'Dinosaur Hunter Seeks the Sound of Nessie', *Scotsman* (Edinburgh), 20 March (2000).
- SCHEIDER, W. & WALLIS, P., 'An Alternate Method of Calculating the Population Density of Monsters in Loch Ness', *Limnology and Oceanography*, 18 (No. 2; March): 343 (1973).

- SCI-NEWS EDITORS, 'Paleontologists Solve Mystery of 'Tully Monster", *Sci-News*, http://www.sci-news.com/paleontology/mystery-tully-monster-tullimonstrum-gregarium-03708.html 17 March (2016).
- SCOTT, Peter, 'Why I Believe in the Loch Ness Monster', *Wildlife*, 18 (March): 110-111, 120-121 (1976).
- SCOTT, Peter & RINES, Robert, 'Naming the Loch Ness Monster', *Nature*, 258 (11 December): 466-468 (1975).
- SHELDON, R.W. & KERR, S.R., 'The Population Density of Monsters in Loch Ness', *Limnology and Oceanography*, 17: 796-798 (1972).
- SHIELS, Doc, 'Mother Nature's Jumbo Jet', *Fortean Times*, No. 42 (autumn): 62-67 (1984).
- SHIELS, Tony 'Doc', *Monstrum! A Wizard's Tale* (Fortean Tomes: London, 1990).
- SHUKER, Karl P.N., *In Search of Prehistoric Survivors: Do Giant 'Extinct' Creatures Still Exist?* (Blandford Press: London, 1995); *The Beasts That Hide From Man: Seeking the World's Last Undiscovered Animals* (Paraview Press: New York, 2003); *Dinosaurs and Other Prehistoric Animals on Stamps: A Worldwide Catalogue* (CFZ Press: Bideford, 2008); *Star Steeds and Other Dreams: The Collected Poems* (CFZ Press: Bideford, 2009); *Karl Shuker's Alien Zoo: From the Pages of Fortean Times* (CFZ Press: Bideford, 2010); *A Manifestation of Monsters: Examining the (Un)Usual Suspects* (Anomalist Books: San Antonio, 2015); *More Star Steeds and Other Dreams: The Collected Poems – 2015 Expanded Edition* (Fortean Words: Bideford, 2015).
- SIEVEKING, Paul, 'Pachyderm in the Loch', *Fortean Times*, No. 210 (June): 10-11 (2006).
- SITWELL, Nigel, The Loch Ness Monster Evidence', *Wildlife*, 18 (March): 102-109 (1976).
- SMITH, Adam S., 'Why Did Elasmosaurids Have Such a Long Neck?', *Plesiosaur Bites*, http://plesiosauria.com/news/index.php/why-did-elasmosaurids-have-such-a-long-neck/ 23 November (2014).
- SMITH, Andrew B., *et al.*, 'Sea-Level Change and Rock-Record Bias in the Cretaceous: A Problem For Extinction and Biodiversity Studies', *Paleobiology*, 27 (spring): 241-253 (2001).
- SMITH, Arthur, 'Nessie, Is This REALLY You?', *Daily Mirror* (London), 11 December (1975).
- SUTHERLAND, Elizabeth, *A Guide to the Pictish Stones* (Birlinn: Edinburgh, 1987).
- SWAIN, Erik, 'N.J. Man Debunks Debunker of 'Nessie", *Sunday Express* (New Jersey), 28 May (1995).
- TAYLOR, Darren, 'Loch Ness Mystery Solved! [re manta identity theory]', *Star* (High Wycombe), 28 October (1988).
- TAYLOR, Michael A., 'Plesiosaurs - Rigging and Ballasting', *Nature*, 290 (23 April): 628-629 (1981); 'Lifestyle of Plesiosaurs', *Nature*, 319 (16 January): 179 (1986); 'Stomach Stones For Feeding or Buoyancy? The Occurrence and Function of Gastroliths in Marine Tetrapods', *Philosophical Transactions of the Royal Society B*, 341 (29 July): 163-175 (1993).
- THESIGER, W., 'The Sea Serpent in the Highlands', *Times* (London), 6 March (1856).

- THOMAS, Charles, 'The "Monster" Episode in Adomnan's Life of St. Columba', *Cryptozoology*, 7: 38-45 (1988).
- TUCKER, Denys W., 'The Zoologist's Tale', *In:* WITCHELL, Nicholas, *The Loch Ness Story* (3rd edit.) (Corgi Books: London, 1989): 214-229.
- VAVREK, Matthew J., *et al.*, 'Arctic Plesiosaurs From the Lower Cretaceous of Melville Island, Nunavut, Canada', *Cretaceous Research*, 50: 273-281 (2014).
- WALSH, Stig A. & NAISH, Darren, 'Fossil Seals From Late Neogene Deposits in South America: A New Pinniped (Carnivora, Mammalia) Assemblage From Chile', *Palaeontology*, 45: 821-842 (2002).
- WATKINS-PITCHFORD, Denys J., *September Road to Caithness and the Western Sea* (Nicholas Kaye: London, 1962).#
- WATSON, Roland, 'Is There Enough Food For Nessie?', *Loch Ness Monster*, http://lochnessmystery.blogspot.co.uk/2012/02/is-there-enough-food-for-nessie_12.html 12 February (2012); 'A Rediscovered Diver's Tale', *Loch Ness Monster*, http://lochnessmystery.blogspot.co.uk/2013/12/a-rediscovered-divers-tale.html 8 December (2013).
- WILEY, John P., 'Cameras, Sonar Close In On Denizen Of Loch Ness', *Smithsonian Magazine*, 7 (June): 96-102 (1976).
- WITZKE, Brian J., 'The Age of Dinosaurs in Iowa', *Iowa Geology*, No. 26: 2–7 (2001).
- WOODLEY, Michael A., *In the Wake of Bernard Heuvelmans: An Introduction to the History and Future of Sea Serpent Classification* (CFZ Press: Bideford, 2008).
- WOODLEY, Michael A., NAISH, Darren, & SHANAHAN, Hugh P., 'How Many Extant Species of Pinniped Remain to be Described?', *Historical Biology*, 20(4): 225-235 (2009).
- YOUNG, Noel, 'Loch Ness Monster: The Unsolved Mysteries of Collisions With Nessie on the Loch', *Scotland Now* [in *Daily Record*], http://www.scotlandnow.dailyrecord.co.uk/lifestyle/loch-ness-monster-unsolved-mysteries-5008548 25 January (2015).
- YOUNG WORLD PRODUCTIONS, *Stranger Than People* (Young World Productions: London, 1968).
- ZAMMIT, Maria, *et al.*, 'Elasmosaur (Reptilia: Sauropterygia) Neck Flexibility: Implications for Feeding Strategies', *Comparative Biochemistry and Physiology, Part A*, 150: 124-130 (2008).
- ZARZYNSKI, Joseph W., '"Seileag": The Unknown Animal(s) of Loch Shiel, Scotland', *Cryptozoology*, 3: 50-54 (1984); *Monster Wrecks From Loch Ness and Lake Champlain* (M-Z Information: Wilton, 1986); '"Mr. Loch Ness," Tim Dinsdale, Dies', *Strange Magazine*, No. 2: 33 (1988).
- ZARZYNSKI, Joseph W. & MEANEY, Pat, 'Investigations at Loch Ness and Seven Other Freshwater Scottish Lochs', *Cryptozoology*, 1: 78-82 (1982).

NESSIE ⊕N THE NET
- A CURRENT LISTING ⊕F LN𝔪-THE𝔪ED
WEBSITES AND BL⊕GS

Loch Ness Monster – *Crystalinks* entry
http://www.crystalinks.com/lochnessmonster.html

Loch Ness Monster – *Wikipedia* entry
http://en.wikipedia.org/wiki/Loch_Ness_Monster

Genuine Facts About "Nessie", the Loch Ness "Monster"
http://www.henryhbauer.homestead.com/lochnessfacts.html

Loch Ness Monster Evidence Website
http://www.loch-ness.com/

Loch Ness Information Website
http://www.loch-ness.org/

Loch Ness Monster
http://www.loch-ness-monster.com/

Loch Ness Monster Nessie Web Site and Live Cams
http://www.lochness.co.uk/

Loch Ness Centre & Exhibition
http://www.lochness.com/

Loch Ness Investigation
http://www.lochnessinvestigation.com/index.html

Loch Ness Monster
http://www.lochnessmystery.blogspot.co.uk/

Loch Ness & Morar Project
http://www.lochnessproject.org/index.htm

The Official Loch Ness Monster Sightings Register
http://www.lochnesssightings.com/

Loch Ness Understood
http://www.lochnessunderstood.com/

Loch Ness Water
http://www.lochnesswater.co.uk/

Loch Ness Monster – 80 Facts to Mark 80 Years Since Nessie was First Spotted
http://www.mirror.co.uk/news/weird-news/loch-ness-monster-80-facts-1826765

Loch Ness Monster Hoaxes – *The Museum of Hoaxes* entry
http://www.museumofhoaxes.com/hoax/archive/display/category/loch_ness_monster/

Government Protection For 'Nessie' – National Archives of Scotland document
http://www.nas.gov.uk/documents/HH-1-588-31.pdf

Legend of Nessie
http://www.nessie.co.uk/

Steve Feltham Monster Hunter
http://www.nessiehunter.co.uk/

Nessieland
http://www.nessieland.co.uk/

The Beast of Loch Ness – Nova Online
http://www.pbs.org/wgbh/nova/lochness/

Nessie's Grotto
http://www.simegen.com/writers/nessie/

Loch Ness "Monster" – *The Skeptic's Dictionary* entry
http://www.skepdic.com/nessie.html

Loch Ness Monster: 50 Fascinating Facts
http://www.telegraph.co.uk/travel/destinations/europe/uk/scotland/11545298/Loch-Ness-Monster-50-fascinating-facts.html

The Legend of Nessie
http://www.thefrasers.com/nessie/

The Loch Ness Giant Salamander
http://www.thelochnessgiantsalamander.blogspot.co.uk/

The Man Who Filmed Nessie – Tim Dinsdale and the Enigma of Loch Ness
http://www.themanwhofilmednessie.com/

INDEX ⊕F ANIΠAL NAΠES

251

AB⊕UT THE AUTH⊕R

Born and still living in the West Midlands, England, Dr Karl P.N. Shuker graduated from the University of Leeds with a Bachelor of Science (Honours) degree in pure zoology, and from the University of Birmingham with a Doctor of Philosophy degree in zoology and comparative physiology. He now works full-time as a freelance zoological consultant to the media, and as a prolific published writer.

Dr Shuker is currently the author of 24 books and hundreds of articles, principally on animal-related subjects, with an especial interest in cryptozoology and animal mythology, on which he is an internationally-recognised authority, but also including a poetry volume. In addition, he has acted as consultant for several major multi-contributor volumes as well as for the world-renowned *Guinness Book of Records/Guinness World Records* (he is currently its Senior Consultant for its Life Sciences section); and he has compiled questions for the BBC's long-running cerebral quiz *Mastermind*. He is also the editor of the *Journal of Cryptozoology*, the world's only existing peer-reviewed scientific journal devoted to mystery animals.

Dr Shuker has travelled the world in the course of his researches and writings, and has appeared regularly on television and radio. Aside from work, his diverse range of interests include motorbikes, the life and career of James Dean, collecting masquerade and carnival masks, quizzes, philately, poetry, travel, world mythology, and the history of animation.

He is a Scientific Fellow of the prestigious Zoological Society of London, a Fellow of the Royal Entomological Society, a Member of the International Society of Cryptozoology and other wildlife-related organisations, he is Cryptozoology Consultant to the Centre for Fortean Zoology, and is also a Member of the Society of Authors.

Dr Shuker's personal website can be accessed at http://www.karlshuker.com and his mystery animals blog, *ShukerNature*, can be accessed at http://www.karlshuker.blogspot.com

His *Star Steeds* poetry blog can be accessed at http://starsteeds.blogspot.com and his *Eclectarium of Doctor Shuker* blog can be accessed at http://eclectariumshuker.blogspot.com

There is also an entry for Dr Shuker in the online encyclopedia Wikipedia at http://en.wikipedia.org/wiki/Karl_Shuker and a Like (fan) page on Facebook.

AUTH⊕R BIBLI⊕GRAPHY

Mystery Cats of the World: From Blue Tigers To Exmoor Beasts (Robert Hale: London, 1989)

Extraordinary Animals Worldwide (Robert Hale: London, 1991)

The Lost Ark: New and Rediscovered Animals of the 20th Century (HarperCollins: London, 1993)

Dragons: A Natural History (Aurum: London/Simon and Schuster: New York, 1995; republished Taschen: Cologne, 2006)

In Search of Prehistoric Survivors: Do Giant 'Extinct' Creatures Still Exist? (Blandford: London, 1995)

The Unexplained: An Illustrated Guide to the World's Natural and Paranormal Mysteries (Carlton: London/JG Press: North Dighton, 1996; republished Carlton: London, 2002)

From Flying Toads To Snakes With Wings: From the Pages of FATE Magazine (Llewellyn: St Paul, 1997; republished Bounty: London, 2005)

Mysteries of Planet Earth: An Encyclopedia of the Inexplicable (Carlton: London, 1999)

The Hidden Powers of Animals: Uncovering the Secrets of Nature (Reader's Digest: Pleasantville/Marshall Editions: London, 2001)

The New Zoo: New and Rediscovered Animals of the Twentieth Century [fully-updated, greatly-expanded, new edition of *The Lost Ark*] (House of Stratus Ltd: Thirsk, UK/House of Stratus Inc: Poughkeepsie, USA, 2002)

The Beasts That Hide From Man: Seeking the World's Last Undiscovered Animals (Paraview: New York, 2003)

Extraordinary Animals Revisited: From Singing Dogs To Serpent Kings (CFZ Press: Bideford, 2007)

Dr Shuker's Casebook: In Pursuit of Marvels and Mysteries (CFZ Press: Bideford, 2008)

Dinosaurs and Other Prehistoric Animals on Stamps: A Worldwide Catalogue (CFZ Press: Bideford, 2008)

Star Steeds and Other Dreams: The Collected Poems (CFZ Press: Bideford, 2009)

Karl Shuker's Alien Zoo: From the Pages of Fortean Times (CFZ Press: Bideford, 2010)

The Encyclopaedia of New and Rediscovered Animals: From The Lost Ark to The New Zoo – and Beyond (Coachwhip Publications: Landisville, 2012)

Cats of Magic, Mythology, and Mystery: A Feline Phantasmagoria (CFZ Press: Bideford, 2012)

Mirabilis: A Carnival of Cryptozoology and Unnatural History (Anomalist Books: San Antonio, 2013)

Dragons in Zoology, Cryptozoology, and Culture (Coachwhip Publications: Greenville, 2013)

The Menagerie of Marvels: A Third Compendium of Extraordinary Animals (CFZ Press: Bideford, 2014)

A Manifestation of Monsters: Examining the (Un)Usual Suspects (Anomalist Books: San Antonio, 2015)

More Star Steeds and Other Dreams: The Collected Poems – 2015 Expanded Edition (Fortean Words: Bideford, 2015).

Here's Nessie! A Monstrous Compendium From Loch Ness (CFZ Press: Bideford, 2016)

Consultant and also Contributor

Man and Beast (Reader's Digest: Pleasantville, New York, 1993)

Secrets of the Natural World (Reader's Digest: Pleasantville, New York, 1993)

Almanac of the Uncanny (Reader's Digest: Surry Hills, Australia, 1995)

The Guinness Book of Records/Guinness World Records 1998-present day (Guinness: London, 1997- present day)

Consultant

Monsters (Lorenz: London, 2001)

Contributor

Of Monsters and Miracles CD-ROM (Croydon Museum/Interactive Designs: Oxton, 1995)

Fortean Times Weird Year 1996 (John Brown Publishing: London, 1996)

Mysteries of the Deep (Llewellyn: St Paul, 1998)

Guinness Amazing Future (Guinness: London, 1999)

The Earth (Channel 4 Books: London, 2000)

Mysteries and Monsters of the Sea (Gramercy: New York, 2001)

Chambers Dictionary of the Unexplained (Chambers: Edinburgh, 2007)

Chambers Myths and Mysteries (Chambers: Edinburgh, 2008)

The Fortean Times Paranormal Handbook (Dennis Publishing: London, 2009)

Plus numerous contributions to the annual *CFZ Yearbook* series of volumes.

Editor

The *Journal of Cryptozoology* (CFZ Press: Bideford, 2012-present day)

STILL ON THE TRACK OF UNKNOWN ANIMALS

The Centre for Fortean Zoology, or CFZ, is a non profit-making organisation founded in 1992 with the aim of being a clearing house for information, and coordinating research into mystery animals around the world.

We also study out of place animals, rare and aberrant animal behaviour, and Zooform Phenomena; little-understood "things" that appear to be animals, but which are in fact nothing of the sort, and not even alive (at least in the way we understand the term).

Not only are we the biggest organisation of our type in the world, but - or so we like to think - we are the best. We are certainly the only truly global cryptozoological research organisation, and we carry out our investigations using a strictly scientific set of guidelines. We are expanding all the time and looking to recruit new members to help us in our research into mysterious animals and strange creatures across the globe.

Why should you join us? Because, if you are genuinely interested in trying to solve the last great mysteries of Mother Nature, there is nobody better than us with whom to do it.

Members get a four-issue subscription to our journal *Animals & Men.* Each issue contains nearly 100 pages packed with news, articles, letters, research papers, field reports, and even a gossip column! The magazine is Royal Octavo in format with a full colour cover. You also have access to one of the world's largest collections of resource material dealing with cryptozoology and allied disciplines, and people from the CFZ membership regularly take part in fieldwork and expeditions around the world.

The CFZ is managed by a three-man board of trustees, with a non-profit making trust registered with HM Government Stamp Office. The board of trustees is supported by a Permanent Directorate of full and part-time staff, and advised by a Consultancy Board of specialists - many of whom are world-renowned experts in their particular field. We have regional representatives across the UK, the USA, and many other parts of the world, and are affiliated with

You'll find that the people at the CFZ are friendly and approachable. We have a thriving forum on the website which is the hub of an ever-growing electronic community. You will soon find your feet. Many members of the CFZ Permanent Directorate started off as ordinary members, and now work full-time chasing monsters around the world.

Write to us, e-mail us, or telephone us. The list of future projects on the website is not exhaustive. If you have a good idea for an investigation, please tell us. We may well be able to help.

We are always looking for volunteers to join us. If you see a project that interests you, do not hesitate to get in touch with us. Under certain circumstances we can help provide funding for your trip. If you look on the future projects section of the website, you can see some of the projects that we have pencilled in for the next few years.

In 2003 and 2004 we sent three-man expeditions to Sumatra looking for Orang-Pendek - a semi-legendary bipedal ape. The same three went to Mongolia in 2005. All three members started off merely subscribers to the CFZ magazine. Next time it could be you!

We have no magic sources of income. All our funds come from donations, membership fees, and sales of our publications and merchandise. We are always looking for corporate sponsorship, and other sources of revenue. If you have any ideas for fund-raising please let us know. However, unlike other cryptozoological organisations in the past, we do not live in an intellectual ivory tower. We are not afraid to get our hands dirty, and furthermore we are not one of those organisations where the membership have to raise money so that a privileged few can go on expensive foreign trips. Our research teams, both in the UK and abroad, consist of a mixture of experienced and inexperienced personnel. We are truly a community, and work on the premise that the benefits of CFZ membership are open to all.

Reports of our investigations are published on our website as soon as they are available. Preliminary reports are posted within days of the project finishing.

Each year we publish a 200 page yearbook

We have a thriving YouTube channel, CFZtv, which has well over two hundred self-made documentaries, lecture appearances, and episodes of our monthly webTV show. We have a daily online magazine, which has over a million hits each year.

Each year since 2000 we have held our annual convention - the Weird Weekend. It is three days of lectures, workshops, and excursions. But most importantly it is a chance for members of the CFZ to meet each other, and to talk with the members of the permanent directorate in a relaxed and informal setting and preferably with a pint of beer in one hand. Since 2006 - the Weird Weekend has been bigger and better and held on the third weekend in August in the idyllic rural location of Woolsery in North Devon.

Since relocating to North Devon in 2005 we have become ever more closely involved with other community organisations, and we hope that this trend will continue. We have also worked closely with Police Forces across the UK as consultants for animal mutilation cases, and we intend to forge closer links with the coastguard and other community services. We want to work closely with those who regularly travel into the Bristol Channel, so that if the recent trend of exotic animal visitors to our coastal waters continues, we can be out there as soon as possible.

Apart from having been the only Fortean Zoological organisation in the world to have consistently published material on all aspects of the subject for over a decade, we have achieved the following concrete results:

- Disproved the myth relating to the headless so-called sea-serpent carcass of Durgan beach in Cornwall 1975
- Disproved the story

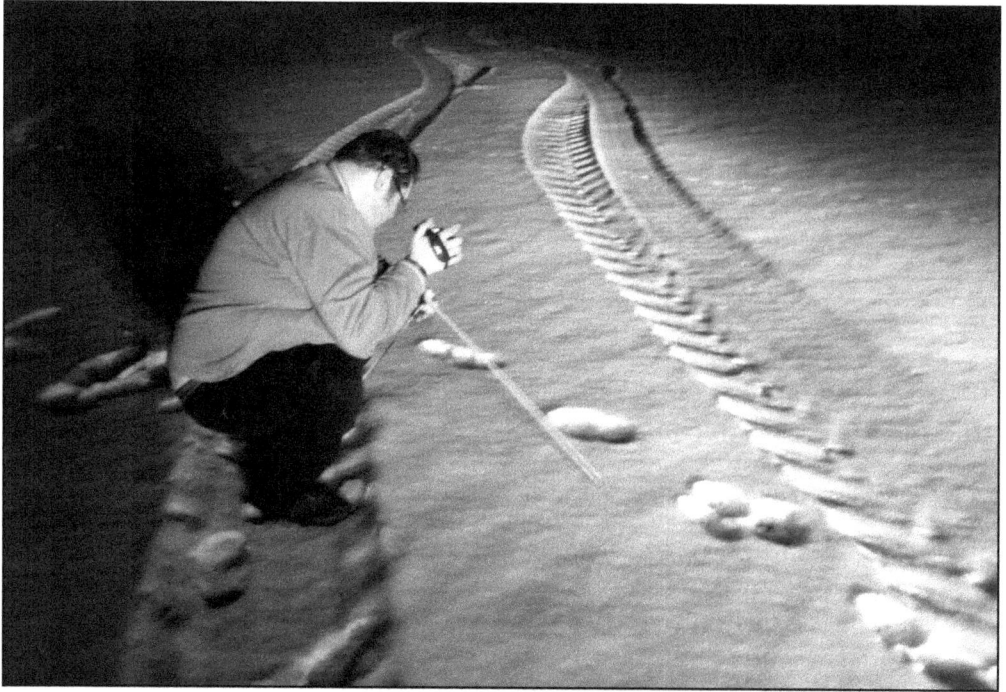

of the 1988 puma skull of Lustleigh Cleave

- Carried out the only in-depth research ever into the mythos of the Cornish Owlman.
- Made the first records of a tropical species of lamprey
- Made the first records of a luminous cave gnat larva in Thailand
- Discovered a possible new species of British mammal - the beech marten
- In 1994-6 carried out the first archival fortean zoological survey of Hong Kong
- In the year 2000, CFZ theories were confirmed when a new species of lizard was added to the British List
- Identified the monster of Martin Mere in Lancashire as a giant wels catfish
- Expanded the known range of Armitage's skink in the Gambia by 80%
- Obtained photographic evidence of the remains of Europe's largest known pike
- Carried out the first ever in-depth study of the ninki-nanka
- Carried out the first attempt to breed Puerto Rican cave snails in captivity
- Were the first European explorers to visit the `lost valley` in Sumatra
- Published the first ever evidence for a new tribe of pygmies in Guyana
- Published the first evidence for a new species of caiman in Guyana

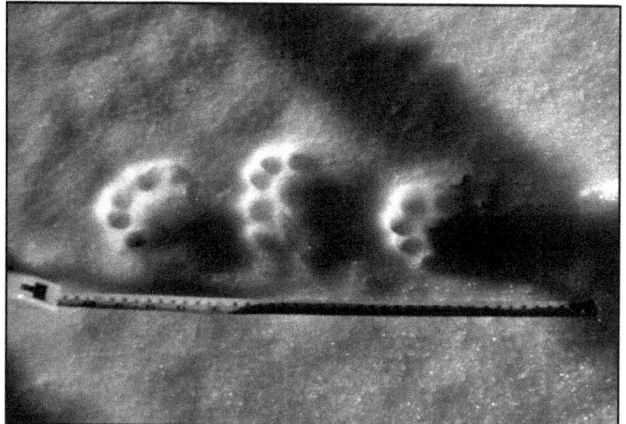

on a monster-haunted lake in Ireland for the first time

- Had a sighting of orang pendek in Sumatra in 2009
- Found leopard hair, subsequently identified by DNA analysis, from rural North Devon in 2010
- Brought back hairs which appear to be from an unknown primate in Sumatra
- Published some of the best evidence ever for the almasty in southern Russia

CFZ Expeditions and Investigations include:

- 1998 Puerto Rico, Florida, Mexico (Chupacabras)
- 1999 Nevada (Bigfoot)
- 2000 Thailand (Naga)
- 2002 Martin Mere (Giant catfish)
- 2002 Cleveland (Wallaby mutilation)
- 2003 Bolam Lake (BHM Reports)

- 2003 Sumatra (Orang Pendek)
- 2003 Texas (Bigfoot; giant snapping turtles)
- 2004 Sumatra (Orang Pendek; cigau, a sabre-toothed cat)
- 2004 Illinois (Black panthers; cicada swarm)
- 2004 Texas (Mystery blue dog)
- Loch Morar (Monster)
- 2004 Puerto Rico (Chupacabras; carnivorous cave snails)
- 2005 Belize (Affiliate expedition for hairy dwarfs)
- 2005 Loch Ness (Monster)
- 2005 Mongolia (Allghoi Khorkhoi aka Mongolian death worm)

- 2006 Gambia (Gambo - Gambian sea monster , Ninki Nanka and Armitage's skink
- 2006 Llangorse Lake (Giant pike, giant eels)
- 2006 Windermere (Giant eels)
- 2007 Coniston Water (Giant eels)
- 2007 Guyana (Giant anaconda, didi, water tiger)
- 2008 Russia (Almasty)
- 2009 Sumatra (Orang pendek)
- 2009 Republic of Ireland (Lake Monster)
- 2010 Texas (Blue Dogs)
- 2010 India (Mande Burung)
- 2011 Sumatra (Orang-pendek)
- 2012 Sumatra (Orang Pendek)
- 2014 Tasmania (Thylacine)
- 2015 Tasmania (Thylacine)
- 2016 Tasmania (Thylacine)

For details of current membership fees, current expeditions and investigations, and voluntary posts within the CFZ that need your help, please do not hesitate to contact us.

The Centre for Fortean Zoology,
Myrtle Cottage,
Woolfardisworthy,
Bideford, North Devon
EX39 5QR

Telephone 01237 431413
Fax+44 (0)7006-074-925
eMail info@cfz.org.uk

Websites:

www.cfz.org.uk
www.weirdweekend.org

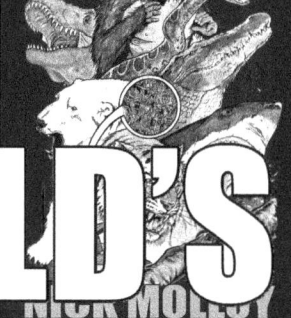

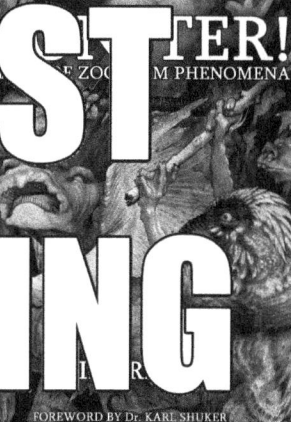
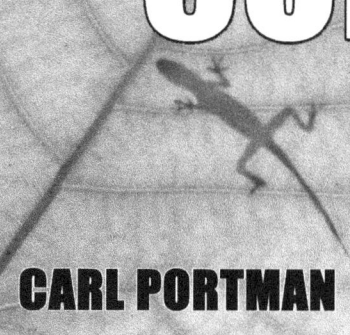
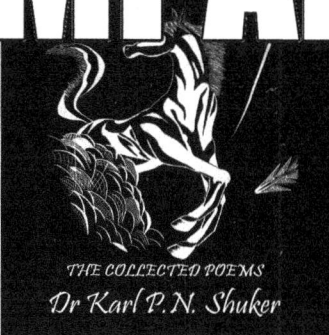

THE WORLD'S WEIRDEST PUBLISHING COMPANY

HOW TO START A PUBLISHING EMPIRE

Unlike most mainstream publishers, we have a non-commercial remit, and our mission statement claims that "we publish books because they deserve to be published, not because we think that we can make money out of them". Our motto is the Latin Tag *Pro bona causa facimus* (we do it for good reason), a slogan taken from a children's book *The Case of the Silver Egg* by the late Desmond Skirrow.

WIKIPEDIA: "The first book published was in 1988. *Take this Brother may it Serve you Well* was a guide to Beatles bootlegs by Jonathan Downes. It sold quite well, but was hampered by very poor production values, being photocopied, and held together by a plastic clip binder.

In 1988 A5 clip binders were hard to get hold of, so the publishers took A4 binders and cut them in half with a hacksaw. It now reaches surprisingly high prices second hand.

The production quality improved slightly over the years, and after 1999 all the books produced were ringbound with laminated colour covers. In 2004, however, they signed an agreement with Lightning Source, and all books are now produced perfect bound, with full colour covers."

Until 2010 all our books, the majority of which are/were on the subject of mystery animals and allied disciplines, were published by `CFZ Press`, the publishing arm of the Centre for Fortean Zoology (CFZ), and we urged our readers and followers to draw a discreet veil over the books that we published that were completely off topic to the CFZ.

However, in 2010 we decided that enough was enough and launched a second imprint, `Fortean Words` which aims to cover a wide range of non animal-related esoteric subjects. Other imprints will be launched as and when we feel like it, however the basic ethos of the company remains the same: Our job is to publish books and magazines that we feel are worth publishing, whether or not they are going to sell. Money is, after all - as my dear old Mama once told me - a rather vulgar subject, and she would be rolling in her grave if she thought that her eldest son was somehow in `trade`.

Luckily, so far our tastes have turned out not to be that rarified after all, and we have sold far more books than anyone ever thought that we would, so there is a moral in there somewhere…

Jon Downes,
Woolsery, North Devon
July 2010

CFZ PRESS

CFZ Press is our flagship imprint, featuring a wide range of intelligently written and lavishly illustrated books on cryptozoology and the quirkier aspects of Natural History.

TRADE MARK

BEWARE OF IMITATIONS

CFZ CLASSICS

CFZ Classics is a new venture for us. There are many seminal works that are either unavailable today, or not available with the production values which we would like to see. So, following the old adage that if you want to get something done do it yourself, this is exactly what we have done.

Desiderius Erasmus Roterodamus (b. October 18th 1466, d. July 2nd 1536) said: "When I have a little money, I buy books; and if I have any left, I buy food and clothes," and we are much the same. Only, we are in the lucky position of being able to share our books with the wider world. CFZ Classics is a conduit through which we cannot just re-issue titles which we feel still have much to offer the cryptozoological and Fortean research communities of the 21st Century, but we are adding footnotes, supplementary essays, and other material where we deem it appropriate.

http://www.cfzpublishing.co.uk/

Fortean Words is a new venture for us. The F in CFZ stands for "Fortean", after the pioneering researcher into anomalous phenomena, Charles Fort. Our Fortean Words imprint covers a whole spectrum of arcane subjects from UFOs and the paranormal to folklore and urban legends. Our authors include such Fortean luminaries as Nick Redfern, Andy Roberts, and Paul Screeton. . New authors tackling new subjects will always be encouraged, and we hope that our books will continue to be as ground-breaking and popular as ever.

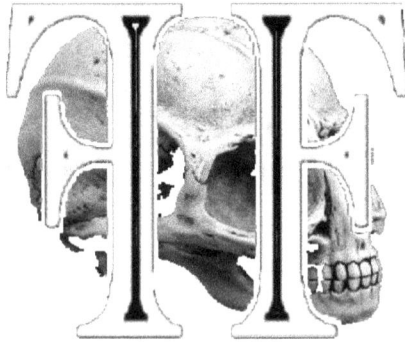

Just before Christmas 2011, we launched our third imprint, this time dedicated to - let's see if you guessed it from the title - fictional books with a Fortean or cryptozoological theme. We have published a few fictional books in the past, but now think that because of our rising reputation as publishers of quality Forteana, that a dedicated fiction imprint was the order of the day.

http://www.cfzpublishing.co.uk/

www.ingramcontent.com/pod-product-compliance
Lightning Source LLC
Chambersburg PA
CBHW062205270326
41930CB00009B/1656